· PALACIVM ·   DVCIS ·

The World of
Titian

TIME
LIFE
BOOKS
®

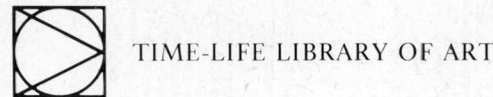

TIME-LIFE LIBRARY OF ART

The World of Titian

c. 1488 – 1576

by Jay Williams
and
the Editors of TIME-LIFE BOOKS

TIME-LIFE BOOKS, New York

About the Author

Jay Williams has published half a dozen historical novels, imaginatively and authoritatively re-creating Biblical times, medieval France, Jacobean Scotland, the crusades of Richard the Lion Hearted, the world of Robin Hood and life in Renaissance Italy. He has written some 20 books for young readers, mostly on scientific subjects but also on art and medieval history. In his novel *The Forger* he revealed a sensitive awareness of the contemporary art scene, a painter's technique and the study of art history—his chief character made a living by forging old master paintings. Mr. Williams lives with his family in England.

The Consulting Editor

H. W. Janson is Professor of Fine Arts at New York University, where he is also Chairman of the Department of Fine Arts at Washington Square College. Among his numerous publications are his *History of Art* and *The Sculpture of Donatello*.

The Consultant for This Book

Jane Costello, Professor of Fine Arts at New York University, is the co-author with Walter Friedlander of *The Drawings of Nicolas Poussin,* the author of *An Outline of the History of Art* and numerous scholarly articles. A former Guggenheim Fellow, Dr. Costello has lectured at Harvard University, The Metropolitan Museum of Art, the University of Pennsylvania and elsewhere. She has taught two series of art courses on television, and was a consultant for *The World of Rubens* in the Time-Life Library of Art.

On the Slipcase

This detail is from Titian's magnificent portrait of a lady at the court of Urbino, which is reproduced in full on page 97.

End Papers

These 16th and 17th Century engravings picture two entertaining aspects of Venetian life.
Front: A Turkish acrobat climbs a taut wire stretched over the Piazza San Marco.
Back: Gentlemen and courtesans take to their gondolas on hot summer nights for music, *al fresco* dining and a cool ride on the canals and lagoon.

TIME-LIFE BOOKS

EDITOR
Maitland A. Edey
EXECUTIVE EDITOR
Jerry Korn
TEXT DIRECTOR   ART DIRECTOR
Martin Mann   Sheldon Cotler
CHIEF OF RESEARCH
Beatrice T. Dobie
PICTURE EDITOR
Robert G. Mason
Assistant Text Directors:
Harold C. Field, Ogden Tanner
Assistant Art Director:
Arnold C. Holeywell
Assistant Chief of Research:
Martha Turner

PUBLISHER
Rhett Austell
General Manager: Joseph C. Hazen Jr.
Managing Director—International: Walter C. Rohrer
Planning Director: John P. Sousa III
Circulation Director: Joan D. Manley
Marketing Director: Carter Smith
Business Manager: John D. McSweeney
Publishing Board: Nicholas Benton,
Louis Bronzo, James Wendell Forbes

TIME-LIFE LIBRARY OF ART

SERIES EDITOR: Robert Morton
Associate Editor: Diana Hirsh
Editorial Staff for *The World of Titian*
Text Editor: Betsy Frankel
Picture Editor: Kathleen Shortall
Designer: Paul Jensen
Assistant Designer: Leonard Wolfe
Staff Writers: Gerald Simons, Peter Yerkes
Chief Researcher: Martha Turner
Researchers: Muriel Clarke, Adrian Condon,
Susan Jonas, Martha Robson, Kelly Tasker
Art Assistant: Nancy Earle

EDITORIAL PRODUCTION
Color Director: Robert L. Young
Assistant: James J. Cox
Copy Staff: Marian Gordon Goldman,
Laurie LaMartine, Florence Keith
Picture Department: Dolores A. Littles, Suzanne Jaffe
Traffic: Arthur A. Goldberger

The following individuals and departments of Time Inc. helped to produce this book: LIFE Staff photographers Larry Burrows, Dmitri Kessel; the Chief of the LIFE Picture Library, Doris O'Neil; the Chief of the Time Inc. Bureau of Editorial Reference, Peter Draz; the Chief of the TIME-LIFE News Service, Richard M. Clurman; Correspondents Maria Vincenza Aloisi, Joan Dupont (Paris); Jean Bratton (Madrid); Margot Hapgood (London); Joe Harriss (Brussels); Elisabeth Kraemer (Bonn); Traudl Lessing (Vienna); Ann Natanson (Rome).

Contents

6

I

Child of
the Renaissance

The village of Pieve di Cadore hangs like an eagle's eyrie among the towering peaks of the Italian Alps. Craggy paths lead from it to a cold blue lake whose waters join the Piave River to rush downstream through precipitous hills and empty into the lagoon of Venice, some 70 miles away. In the 15th Century, this region of Cadore and its tough, loyal people were part of the rampart that stood between the Venetian republic and its German neighbor to the north, that loose collection of states called the Holy Roman Empire. From Cadore's forests, lumber floated downriver to Venetian shipyards in the *Zattere,* the "Quarter of the Rafts," to be made into the swift galleys that gave Venice its power on the sea. Unexpectedly, Cadore also gave Venice another treasure, far richer than either ships or allies, when some time toward the close of the 15th Century Tiziano Vecellio was born.

As Titian, the most versatile and productive painter of his day, Tiziano Vecellio was idolized by common folk and mighty princes of the state and church. For almost a century he stood at the very top of his profession, excelled by none as a portraitist, a painter of landscapes and religious scenes. And yet, for all his fame and his familiarity with the powerful courts of Europe, Titian remained essentially a child of the mountains. The ambition that drove him came as much from his ingrained attitudes as it did from the need to find outlets for his talent. Like a peasant whose one concern is to store up grain against the winter, Titian worked for security. Like a peasant he was by turns dogged and as pliable as putty, obsequious and nagging, devious about money matters and deeply rooted in his native place. Also, like a peasant, he was close-mouthed about himself. What is known about his life is known mostly through his effect upon others—the painters who influenced him or were influenced by him, the men and women whom he memorialized on canvas. He was like some great, deep river whose nature can be understood only by the channel it cuts through the earth, through the shapes and colors of the banks and borders that mark its passing.

The date of Titian's birth is uncertain. The records of 15th Century mountain villages were transitory affairs, and birthdates of countryfolk

Titian was at an age when most men are content to retire when he painted this robust self-portrait. (Its exact date is unknown, but the artist was probably in his late sixties.) Shrewd and self-reliant, he had become wealthy and famous through his art, and the strength evident in his powerful hands was to persist through the 20 creative years he still had to live. Wearing the golden chain of knighthood bestowed on him by Emperor Charles V, Titian boldly confronts the world.

Self-portrait, c. 1550-1566

were kept with less care than those of their rulers. Tradition says that he lived to be 99, which would have made his birth date 1477, but there is no evidence for this tradition. In 1571, Titian himself claimed to be 95, making his birth date 1476, but he did so under circumstances that cast some doubt on the claim. He was writing a letter to King Philip II of Spain, pleading for money that the monarch owed him. In referring to himself as "an old man of 95" he may have added a few years to his age in a shrewd attempt to gain sympathy. Most scholars agree that he was probably born in 1488.

The family into which he was born was not rich by any means, but not badly off; the Vecellio were for the most part lawyers or soldiers. They appear to have had business connections in Venice and to have been much respected in their own community, taking an active part in civic affairs. Titian's grandfather was a member of the governing council of the Cadore region; Titian's father was a captain in the local militia and later served as overseer of the town's grain supplies and inspector of mines. Tiziano is said to have been named for a saint in whom the Vecellio family had a proprietary interest. Before his canonization, St. Tiziano had been a bishop, and had officiated from a chapel that ultimately belonged to one of the Vecellio ancestors. Tiziano was a common name among the Vecellio—although artistic talent was not.

It would be interesting to know when Titian's gift for art was first noticed. A mountaineer's life is never easy, and in 15th Century Cadore it must have been an endless struggle. The soil of the region produced barely enough food to last its people a third of the year, and the income from its other resources—lumber, a few iron mines, hides and fleece—went to buy additional grain. It is hard to imagine anyone paying heed to the playful scribblings of a child. The children of Cadore must have been left largely to themselves, to roll like puppies in the sun, climb among the rocks, fish in the mountain streams. But Italians have always cosseted

Toward the end of his life, Titian painted a Madonna and Child that depicts the artist's Seventh Century namesake, St. Tiziano. Dressed as a bishop, the saint kneels before the Virgin; behind him, in a rare and realistic self-portrait, is the aging Titian himself, holding a crosier. St. Andrew, one of the 12 disciples, squats at the right. The work was produced for Titian's family chapel in Pieve di Cadore—the artist's birthplace—where, after nearly 400 years, it is still displayed.

their children, and someone must have been aware of Titian's budding talent. Perhaps he was discovered scrawling with a piece of charcoal on a whitewashed wall. There is a charming story told by a cousin of Titian's in a biography published many years after the painter's death. According to this story, the boy drew attention to himself by painting a portrait of a Madonna on the side of a house, using for his colors the juice pressed from flowers. But apart from a mid-Victorian artist named William Dyce, who painted an unbearably saccharine picture of this improbable event, no one has ever given the tale much credence. However Titian's talent first came to light, evidently he showed promise, for his father decided to send him to Venice to be apprenticed to an artist and to learn the mysteries of the craft. Titian was then about nine years old.

It was a decision made for sound, practical reasons—as sound and practical as those of a modern parent who decides to enter a mathematically gifted son in a college of engineering. To be a painter or a sculptor in the Renaissance was to be a respected and sought-after craftsman. More than that, for a man of really splendid talent, art provided an entrée to the society of princes and to all the rewards that went with princely favor. The painter was not a man living in a rarefied world, bent upon self-expression; he served a practical function. He was a decorator, a recorder of events, real or fancied, a purveyor of beauty for people to whom beauty was an integral part of life. Some men, observed a contemporary Venetian writer named Lodovico Dolce, "rank painting very low . . . . These men do not know how useful, how necessary, how ornamental it is to the world in general." Painters, Lodovico pointed out, enriched and decorated palaces and public buildings, produced maps and views for kings and generals, and illustrated incidents in the life of the Holy Family and the saints that were "of great benefit to the devout Christian in awakening his imagination, fixing his attention, and raising his ideas to heaven."

Although Lodovico did not say so, the painter also enhanced his patron's public image. The lords of the great city-states of Renaissance Italy were the sons of bankers and businessmen who had grown rich on trade. As rivals of the older landed nobility, they were eager to display their wealth, their intellectual pretensions and their courtly manners. Serving these merchant princes were mercenary armies whose captains were often rewarded for their services with gifts of land, of towns and even of whole districts, and who thenceforth became themselves the founders of great dynastic families. In gratitude and pride, these new aristocrats filled their churches with devotional paintings of Madonnas and saints and crucifixions—paintings that often included the figure of the donor kneeling stiffly in the foreground. They memorialized themselves in portraits and in pictures of their great military victories; they embellished their cities with statues of their illustrious ancestors and with splendid marble palaces and public buildings.

There was more to this interest in art than the mere desire for surface glitter. Along with its discovery of the visual glories of ancient Greece and Rome, the Renaissance had also been awakened to the glories of classical learning. The new humanism had sent men back into the works of Aristotle, Horace and Virgil—writers who believed that nature was the measure of all things. Looking about them, men saw new beauty in landscape and in the human form. The concept of individuality once again flowered, gradually replacing the medieval view that man was a fixed unit in a rigidly ordered society. Instead of filling a predetermined role—peasant, artisan, priest or knight—man was now seen to fill the role for which his own particular talents and personality fitted him. And just as the Greeks in their sculpture had conceived of an ideal human form, so the Renaissance conceived of an ideal human deportment. The perfect prince or the perfect courtier was a man accomplished in manners and tastes, skilled in every art, social and intellectual, and able above all to perceive and appreciate beauty.

With these changing ideas, the painter's eye changed too. When the medieval artist painted a Madonna, he was in a sense performing an act of devotion. He approached his task as he might recite a prayer: he followed a strict convention. The highly stylized image he produced in this way was not a person but a religious symbol. The Renaissance artist discarded this approach and painted the Madonna as a woman. His picture portrayed a real person placed against a background of real trees and hills. When he discovered the technique of drawing in perspective, this natural background acquired depth, and the painting miraculously opened a window in a solid wall. And when he acquired the new technique of mixing his pigments with oil, this illusion was intensified, for oil made it possible to paint light and shadow with far greater subtlety, and thus to increase the sense of distance and dimension.

From its origins in Tuscany, and especially in the city of Florence, this concern for reality in art spread throughout Italy. It did not reach Venice until the latter part of the 15th Century, but when it did it touched off an artistic revolution. The painters of Venice took up the style with a gusto typical of their city, but they also gave it a direction that altered the whole future of painting. The nature of this change was part and parcel of the peculiar nature of Venice. Up until about the middle of the 15th Century the most advanced of the Venetian arts was the art of making and spending money. In pursuit of this, Venice had become the most colorful and cosmopolitan city in Europe. Its wharves were piled high with exotic goods from the East, its street noises carried an obbligato of foreign tongues, its merchants were as familiar with the customs and way of life of the Levant as they were with those of Europe. While the painters of Florence were exploring the subtleties of creating three-dimensional forms, the painters of Venice preferred an art that was richly ornamental, a fusion of Byzantine and Gothic.

Then, on the nearby island of Murano, noted for its painting workshops as well as its glass factories, the workshop of an artist named Antonio Vivarini began to turn out altarpieces and devotional paintings in which the figures had a more sculptural quality. Antonio's shop was a family affair, including his brother-in-law and his brother and his son. In addition to softening the contours of their human figures, the Vivarini made some attempt to simplify the intricate draperies of the Gothic style, and occasionally they tried to introduce some perspective into their background. But the new realism in their work generally struggled unsuccessfully against a continuing regard for the older Gothic conventions, against stiff postures, gilded thrones and halos, and elaborately carved wooden frames. It remained for another Venetian painter, a contemporary of Antonio Vivarini, and like him the founder of a dynastic painting family, to break with these conventions.

Jacopo Bellini learned to paint as an apprentice to Gentile da Fabriano, one of the leading painters of the day. In 1408, Gentile was invited to Venice to decorate the walls of the Great Council Hall, the meeting place of the Venetian government; afterwards when he went to Florence, he may have taken Jacopo with him as a student. Jacopo's reputation never equaled that of his two sons, Gentile and Giovanni, but he seems to have been highly regarded and to have accepted commissions in Padua and Ferrara as well as in his native Venice. Giorgio Vasari, the Renaissance artist who chronicled the lives of all the other Renaissance artists, observes that Jacopo was considered to be "the first of his profession" in Venice—although he adds that Jacopo only acquired that reputation after the competition had moved elsewhere.

Among the few surviving examples of Jacopo's work are two sketchbooks, one preserved in the British Museum and the other in the Louvre (*page 18*). They reveal a lively, inquiring mind. Subjects range from knights fighting dragons to ambitious street scenes in which peasants and courtiers mingle in complicated architectural settings of walls, staircases, towers and arches. It is easy to see, in Jacopo's sketches, the conflict between the old style of painting and the new. The mountains have mass, but their swirling terraces are a bit insubstantial, like a blancmange pudding that is about to fall apart; a funeral moves directly toward the viewer in a very daring and modern way, but behind it are little hills in the Gothic tradition, pointed humps crowned by a tower or a single tree.

There are echoes of these curious drawings in the work of Jacopo's sons, but as painters they soon surpassed him. Jacopo did not mind; far from being jealous, he encouraged them constantly and was delighted

One of the world's largest rooms, the Great Council Hall in the Doge's Palace dwarfs a ceremonial gathering of Venetian noblemen in this 18th Century engraving by Brustolon and Canaletto. From the date of its first use in 1423, the chamber housed legislative meetings, elections and banquets. (At one feast 3,000 guests came to dine.) Over the centuries the hall was decorated by Venice's leading artists. Before it was gutted by fire in 1577 the room contained paintings by Gentile da Fabriano, the Bellini brothers, Carpaccio and Titian. Afterward Veronese and Tintoretto were given major commissions; Veronese's *Apotheosis of Venice* may still be seen on the ceiling; Tintoretto's *Paradiso*, which measures 72 by 23 feet and was the largest oil painting ever seen at the time of its unveiling in 1590, covers an end wall.

with their prowess. As for Gentile and Giovanni, they appear to have loved both their father and each other. Each, says Vasari, constantly played down his own work and praised the work of the other, and "thus they modestly sought to emulate each other no less in gentleness and courtesy than in excellence as painters." All three Bellini worked together for a time in the father's studio. The sons served Jacopo first as assistants, then later apparently as partners—since there is some mention of splitting commissions. Gentile, who maintained the studio after Jacopo's death, also inherited his father's style; his work is solid, dignified and decorous, even a little pompous. Giovanni, however, went off on his own to paint in a more naturalistic style. There is a softness, a warmth and a sweetness to his work that is very winning. But it was Gentile who first received official recognition.

In 1469 the Holy Roman Emperor Frederick III made Gentile an honorary knight. Five years later, the city of Venice conferred another honor on him by choosing him to undertake its most important artistic project —the restoration and repair of the paintings on the walls of the Great Council Hall. These paintings had been executed less than a century before by the Veronese artist Pisanello and by Jacopo Bellini's teacher, Gentile da Fabriano, for whom Gentile Bellini may have been named. The paintings, which depicted great moments in Venetian history and reminded the lawmakers of their city's accomplishments, were beginning to deteriorate. Pisanello and Gentile da Fabriano had used fresco, the customary method of decorating walls throughout the Italian peninsula— and fresco, which worked well in other cities, suffered in watery Venice.

In the fresco method, the wall surface was coated with a thin layer of plaster, and the pigments, mixed with water, were applied while the plaster was still fresh; as it dried, the colors were locked into it. The method had its disadvantages. Fresco colors were limited to those pigments that were not altered by the action of the lime in the plaster, and the fresco painter had to work quickly, laying on only enough fresh plaster to complete one small section at a time; to avoid a patchy look, he undercut the edge of each day's work and smoothed the next day's plaster against it. Fresco painting could be surprisingly durable—the colors of some frescoes have remained clear and true for centuries—but its great enemy was dampness, which could, in time, disintegrate the plaster; in sea-girt Venice, that danger was ever present. And so, although frescoes were common, Venetians came to prefer paintings on canvas or wood.

Of the two, canvas eventually became the favorite—for reasons that dovetailed nicely with the Venetian character. Canvas was simply more practical; unlike wood it did not split or crack, and it was far more portable. If a Venetian gentleman wanted to alter the look of a wall, he could remove the canvas with relative ease. Canvas could also be patched when damaged and could be rolled up and transported more cheaply than a painting on wood. Another advantage came to light when oil began to be used in painting: canvas and oil went together like wine and cheese.

Before canvas came to be so popular, most pictures, other than fresco paintings, were done on wood in tempera: the pigments were mixed with a binding medium, usually egg yolk, and then were thinned with water to the right consistency for painting. The surface to be painted was primed with gesso, a mixture of plaster of paris and glue; over this the paint was built up in layers. Because tempera dries so rapidly, the painter made highlights and shadows by stippling his forms with tiny dots or crosshatching them with fine lines. The finished work was protected with a coat of varnish that made it as smooth and glossy as enamel—more like a jewel than a painting. With oil painting on canvas, much of this jewel-like quality was lost. But the new medium, in the hands of imaginative artists, made possible a wholly new kind of art—and the Venetian painters were among its first exponents.

Vasari, the inveterate collector of rumors and anecdotes, tells a fascinating but baseless tale about how oil painting came to Venice. Early in the 15th Century, Vasari writes, the Flemish painter Jan van Eyck invented a way of mixing pigments with boiled linseed or nut oil so that they dried firmly and remained fast; furthermore, the colors had so much life that they needed no varnish to give them luster. One of van Eyck's paintings was seen in Naples by a Sicilian artist named Antonello da Messina, in the collection of the Spanish king who ruled Sicily and Naples. Antonello at once dropped everything and hurried off to Flanders to learn van Eyck's secret. By fawning over the Flemish artist, he got what he was after and returned to Sicily. Eventually Antonello made his way to Venice, "having found in that city a life just suited to his taste," but there became the victim of his own confidence trick. The painter Domenico Veneziano began inviting him everywhere, flattering him and treating him as if Antonello were his oldest and dearest friend. Not wishing to be outdone in courtesy, Antonello shared his precious secret with Domenico. Shortly thereafter, the Sicilian painter died, a victim of pleurisy brought on by the fogs and cold wind of the city of the lagoons, and Domenico was left the sole master of the art of painting in oils.

So goes Vasari's story, but scholars, cross-checking dates and references, are inclined to believe that Domenico never received any instruction from Antonello, and doubt that Antonello ever traveled to Flanders. More to the point, the technique of painting with oil-mixed pigments was probably not the invention of one man, nor even of one time or place. What is more likely is that it came into use slowly and gradually, as historical changes usually come. One painter added to the attempts of another until by degrees the full range of the possibilities of the new medium were understood. There are references to oil-mixed pigments in a treatise on art written in the 10th Century by the Roman painter Eraclius; in an early 12th Century *Catalog of the Various Arts* by the German monk Theophilus; and in *The Book of Art* written in the early 15th Century by the Tuscan painter Cennino Cennini. Oil and tempera may even have been used interchangeably for a time, the tempera being used for delicate skin tones, the oil for the rich fabrics of robes and draperies. For a time, too, artists worked with oil much as they had worked with tempera. The character of early oil paintings is essentially the same as that of tempera, precise and meticulous; only the absence of stippling and crosshatching marks the change in medium.

The effect of the oil-painting technique upon the Bellini brothers was as different as the character of their paintings. Giovanni, the more venturesome of the two, was the first to experiment with it. Sometime around 1475, he painted a *Resurrection* in tempera for the little Church of San Michele on the island of Murano, which he glazed with a thin film of transparent oil color rather than the conventional varnish. From this he went on to heavier oil glazes, which completely altered the character of his paintings, giving them a softness and a luminosity very like that of true oil painting. Giovanni's paintings, in fact, laid the groundwork for the next generation of Venetian painters, who used oil almost exclusively. On Gentile, however, the introduction of oil made very little impression. Although he probably appreciated the richness it gave to colors, oil for Gentile was simply an improvement in method, not a stimulant to a new mode of expression.

Five years after Gentile had been given the job of restoring the frescoes in the Great Council Hall, he persuaded the government to let him produce new paintings instead; possibly he was not very keen on touching up the work of other hands. Before he got very far with this project, however, history intervened. In 1479 the Sultan of Turkey, Mohammed II, asked the Venetian authorities to send him a good painter, one accustomed to doing portraits, since Moslem painters were forbidden by their religion to make images. The Venetian government asked Gentile if he would be willing to go, and Gentile agreed—but like a true Venetian, he made a bargain. He would do so only on condition that his brother Giovanni be given the project for the Great Council Hall, thus keeping the job in the family, and also on condition that Giovanni be given a broker's license in the Fondaco dei Tedeschi, the German merchants' trading center. The value of the license sprang from a civic decree that made it compulsory for foreign merchants who operated in Venice to do business through a local middleman, a "broker" whose function was little more than making introductions. In short, this pleasant sinecure would provide Giovanni with a regular income for doing practically nothing.

Gentile spent a year in Constantinople and was well treated by the Sultan. A man of immense energy and intelligence, Mohammed II had made himself supreme in the Near East and, like many Oriental despots, had become a patron of the arts. Gentile's drawings and paintings, which included several portraits of the Sultan himself, delighted him. Unlike the stylized Byzantine work to which he was accustomed, they seemed to contain the very breath of life. He was also captivated by Gentile's solid and straightforward personality, and often engaged him in conversation. Once, when he asked Gentile's opinion of a much-favored dervish whose portrait he had asked Gentile to paint, Gentile replied, without mincing words, "Sire, he looked to me like a madman." A discussion about another work ended somewhat differently, however. Gentile had brought the Sultan a painting of the severed head of John the Baptist which Mohammed II found not quite to his liking. The neck, he said, was too long for that of a decapitated man: to make his point he ordered a slave brought before him and cut off his head on the spot. After this incident Gentile is said to have thought seriously of going home.

When he finally did return, Venice greeted him with much fanfare. He was commended for his behavior as a cultural ambassador and, in more substantial terms, was granted a handsome annual pension of 200 *scudi* for the rest of his life. Thereafter, Gentile and his brother Giovanni, although they occasionally worked together on large projects, increasingly went their separate ways as painters. Giovanni, continuing his experiments with the new medium of oil, painted gentle golden Madonnas that were remarkable for their humanity; Gentile, secure in the public's esteem, concentrated on the meticulous crowd-filled scenes of Venetian life that delighted its people. In his last decade, Gentile devoted himself to a series of wall paintings for two important Venetian institutions, the Scuola of St. John the Baptist and the Scuola of St. Mark. The *scuole,* or schools, were not institutions of learning but fraternal organizations. Although their purpose was primarily devotional, they also functioned as gentlemen's clubs and mutual assistance leagues. Their members included craftsmen and highly placed patricians, and they met together not only to worship but also to help each other in times of sickness and need.

Gentile's paintings for the Scuola of St. John, the best-known of all his works, illustrate three incidents centering on the Scuola's proudest possession, a fragment of the True Cross. The relic had been given to the Scuola in the last quarter of the 14th Century by one of its wealthy members. Every year, on the feast days of the various saints, it was housed in a magnificent gold casket beneath an embroidered canopy and borne in procession around the city, surrounded by the members of the fraternity carrying lighted candles. On one such occasion the crowd pressed so close to catch a glimpse of the sacred object that its bearers were jostled, and the casket fell into a canal. Legend says, however, that instead of sinking to the bottom, it floated. A number of the brothers leaped into the water to rescue it, but to everyone's astonishment the casket kept slipping away. Finally the Scuola's leading member, the man who was the relic's official guardian, jumped in. Immediately the casket drifted toward him, permitting him to seize it and carry it ashore.

This miraculous rescue is the subject of the second in Gentile's series of three paintings for the Scuola. The others show the relic being carried in a procession through the Piazza San Marco *(page 20)* and being used to heal a sick man. All three paintings are done in a combination of oil and tempera, and all three are notable for their painstaking accuracy. The members of the Scuola are portrayed with careful attention to detail, and the settings are drawn with such architectural exactitude that they seem almost like photographs. One shows the jewel-like Piazza San Marco looking very much as it does today; another shows a second, now-famous Venetian landmark, the hump-backed Rialto Bridge—except that in Gentile's painting the bridge is wood (subsequently it was rebuilt in stone). In their skillful planning and execution these paintings reveal an artist at the height of his power, a master craftsman. And yet the work of Gentile, for all its assurance and finish, was less important to the development of painting than that of his brother Giovanni. In 1497, when Titian came down from Cadore to begin his apprenticeship, Venetian art stood on the threshold of a new way of painting—and the man who led the way was Giovanni Bellini.

The Birth of a Style

Venetian painting evolved its unique character out of a slow succession of artistic influences. The seabound city, facing the east and with her back nestled into a curve of the Italian peninsula, was particularly exposed to the ebb and flow of foreign ideas and images. For centuries after Venice's founding in the Fifth Century, the prevailing mood was Oriental, and the flat, colorful, Byzantine style—expressed in stark, otherworldly Madonnas like the one at right—held sway. Then in late medieval times, from the north, came the expressive, linear Gothic style with its new interest in the natural world. Still later, from the south, especially from Renaissance Florence, came the warming influence of humanism, a burst of interest in classical antiquity and some new solutions to pictorial problems, especially the handling of space and depth through linear perspective.

But Venetian painting in the Renaissance developed as more than merely a sum of imported influences. It also reflected a world stretched between the sea's distant horizon and the dazzling pageant of Venetian palaces and churches; a world seen in the transient effects of fog, mist and the shimmer of light on water.

By the end of the 15th Century, and largely through the work of one family, this synthesis of a Venetian style reached greatness. Jacopo Bellini and his sons Gentile and Giovanni, along with their colleague, Vittore Carpaccio, were Venice's own masters.

This bejeweled Byzantine Madonna is an icon believed to have been brought to Venice after the Crusaders' sack of Constantinople in 1204. Such refined, decorative art left its mark in the Venetian love for luxurious ornamentation.

Nicopeian Madonna, 9th or 10th Century(?)

Jacopo Bellini: *Saint John the Baptist Preaching*, c.1438

Jacopo Bellini was among the first Venetian artists to experiment with the classical principles of painting that had been rediscovered elsewhere in Italy. In his sketch above, the gowned figures are rendered in the flowing, rhythmic lines of the Gothic tradition. Yet the Roman arch, fluted columns, the crisply drawn buildings that recede so precisely into the background, and even the foreshortened horse, reveal the impact of Renaissance ideas.

Like his father, Gentile Bellini struggled to portray three-dimensional objects in a convincing, deep space. In his painting at the right, the trees and winding paths seem as stylized as the clothing in Jacopo's sketch, but the hills, the classical building and the human figures have a certain solidity and massiveness. Still, Gentile had much to learn about relating objects and figures in space and creating a coherent sense of depth within a scene.

Gentile Bellini: *Saint Francis*, c.1465-1480(?)

Gentile Bellini: *Procession of the True Cross in Piazza San Marco*, 1496

Gentile Bellini's great contribution to Venetian painting was the narrative scene. His picture at the left—one of a series—commemorates a healing miracle that occurred in 1444 while a relic of the True Cross was paraded around Piazza San Marco. Gentile has painted a vast panorama of Venetian life, meticulously filled with details of architecture and costume and with recognizable portraits of some of his contemporaries.

Vittore Carpaccio followed Gentile's example when he painted scenes from the life of the medieval Saint Ursula, whose betrothal to a pagan English prince and subsequent martyrdom by Huns was a popular Venetian legend. But compared with Gentile's rather stiff composition, wooden figures and awkwardly defined space, Carpaccio's figures are animated and his picture *(below)* is alive with bright light and color—especially the modulated tones of red and green that weave a bond between the activity in the foreground and the background. The ambitious picture is filled with episodes: the departure of the English prince from his parents at the left, Ursula's farewells at the right and the embarkation of the couple in the right background. Not only a record of Venetian life, Carpaccio's narrative is also a fantasy—even the architecture is a mixture of reality and imagination. But it is truly a Venetian painting—dramatic, colorful and richly detailed. This quality will be seen again and again in works by later artists.

Vittore Carpaccio: *The Bride and Bridegroom Taking Leave of Their Parents,* 1495

The Venetian style, which was to erupt into full glory with Giorgione and Titian, neared maturity in the works of Giovanni Bellini. More highly skilled as a painter than both his father and his more conservative brother, Giovanni created fully three-dimensional figures that exist in airy, well-defined spaces and are incorporated into convincing landscapes. In his masterful *Transfiguration (below)*, for example, a serene Christ is flanked by the revered prophets Moses and Elias, while Peter, James and John fall terrified before Him in truly human awe. The precisely outlined, hard forms of earlier paintings are gone. Giovanni models form with the soft light and true color of the atmosphere. The setting is filled with believable detail, and the viewer's eye is drawn into long vistas of a rural town and pale, distant mountains. In the *Madonna* at the right, Giovanni's control of natural light and color endows the figures and the flesh with a reality that is almost palpable. It is this form-revealing language of light and sensuous color that would become a Venetian trademark.

Giovanni Bellini: *Transfiguration*, c. 1480

·IOANNES·BELLINVS·P·
1487

Giovanni Bellini: *Madonna of the Trees*, 1487

24

II

"The Most Triumphant City"

"Nothing is like it, nothing equal to it," rhapsodized Elizabeth Barrett Browning, but D. H. Lawrence found it an "abhorrent, green, slippery city." Venice has always inspired such violent extremes. To one Italian poet, Torquato Tasso, it was the "ornament of Italian dignity," to another, Betuzzo da Cotignola, a place of "false treachery." The French visitor Philippe de Commynes thought Venice "the most triumphant city I have ever seen," but the British visitor Edward Gibbon remarked that it had given him "some hours of astonishment" followed by "some days of disgust." Even Venetians themselves were not always happy about their city. One ruling Doge, noting the imperfections of its site, suggested that Venice might be better off if it packed up and moved to the shores of the Bosporus. Half land and half water, an architectural wonderland built on mud, Venice was by its very nature a city of contradictions. Even its origins are ambiguous.

The sandy islands within the Venetian lagoon and the long breakwater islands that protect Venice from the sea were originally inhabited by simple fisherfolk. Then, during Roman times, barbarian invaders from the north—Goths and Visigoths—drove people from the mainland onto these islands for safety, to be joined later by other peoples, refugees from Attila the Hun and from the Lombards, who established a kingdom in northern Italy in 568. Venice is supposed to have been founded "around noon" on March 25, 421, by the inland city of Padua as a terminus for its sea trade. But the ardently independent Venetians always denied this vigorously, especially during the years when Padua was their great commercial rival —and since the document containing the date is known to be a forgery, perhaps the Venetians were right. In any case, by the Sixth Century Venice was a tiny, autonomous state, possessed of ships and port facilities fine enough to attract the attention of Belisarius when the great Byzantine general arrived in Italy to free the peninsula from the Goths. In 539, Belisarius asked Venice for help, and Venice agreed—for a price. It was the first of many such political bargains that, in the centuries to come, raised the city to a position of great power in the Mediterranean world.

As their influence grew, so did their watery city. The muddy ground

Giovanni Bellini, noted principally for his religious paintings, was such a fine and popular portraitist that virtually every important Venetian sought to have his picture painted by him. The reason for his success shows through strongly in this portrait of a Doge. Bellini has given the unprepossessing man such an air of quiet authority and simple strength that he becomes almost an archetype of the benign ruler.

Giovanni Bellini: *Portrait of Doge Leonardo Loredano,* c.1501 was shored up with post-and-wattle matting; the islands were bridged; and heavy piles were driven deep to support increasingly massive buildings. Very early, the city acquired a republican form of government whose continuing stability was to be one of its major assets. Beginning with a simple council of magistrates representing each of the Venetian islands, the republic gradually evolved into a complex and carefully organized system of government composed of many levels. At the base was an elective assembly, the Great Council, chosen by the Venetian aristocracy from among its own ranks; at the head was a Doge, or duke, elected by the Council, and his staff of six advisers, the Signoria. Between the Great Council and the Doge was a smaller legislative body, the Pregadi, or senate, and an administrative arm, the Collegio, or cabinet. By the time of the Renaissance, however, the real power in Venice was vested in the Council of Ten, a so-called committee of public safety. This elite group, set up to deal swiftly and effectively with anything that threatened the stability of the state, soon determined the course of everything in Venetian life from public morals to international diplomacy.

The source of Venetian power was, from the start, the Venetian fleet. With three rivers at its back and a sea before it, Venice turned naturally to maritime trade—and quickly brought that trade under state control.

The island city of Venice has almost doubled in size between 1200 and the present day—the result of careful and laboriously slow landfill operations. The outline maps at the right show how the margins of the city have been gradually extended over the centuries. Because Venice is built on the soft subsoil of a lagoon, additions to its borders have to be of a similarly flexible material; masonry underpinnings are impractical because of the strong action of tides through the canals and because constant exposure to weather and temperature changes would crack and destroy them. Wooden pilings driven deep into the sandy subsoil provide virtually the only solid support for Venetian buildings. Venice's position in the lagoon—she is separated from the open Adriatic only by a long strip of sand bars—is both a hazard and a virtue; if the daily flow of salt water to and from the sea was blocked off, the lagoon would silt up and surround the city in an unhealthy, stagnant swamp. The city would also be robbed of her greatest asset— deepwater channels that made her a power in international trade. The lagoon has been described as Venice's lung; although the hazard of flooding is constant, the daily respiration of tides keeps the city breathing healthily.

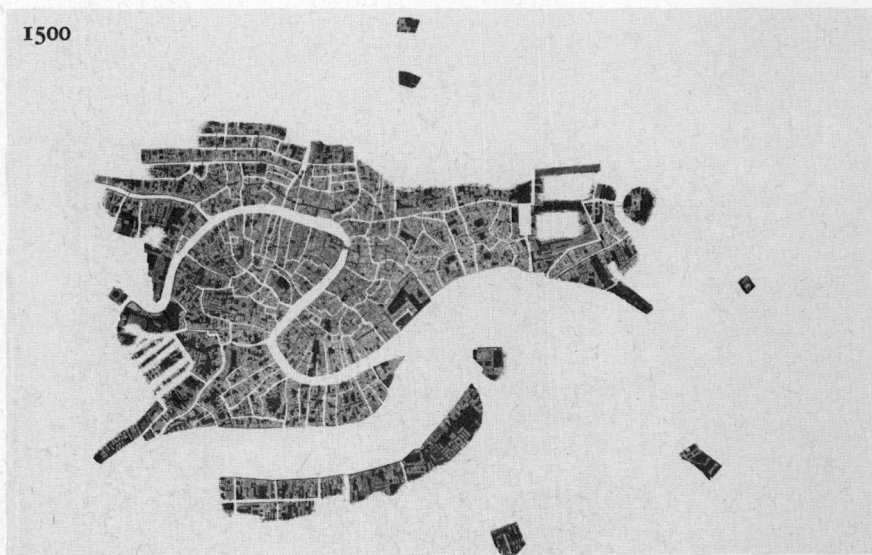

1200

1500

Hundreds of rules regulated the size of Venetian ships, the complement and salaries of their crews, the weight and nature of their cargoes, their destinations and times of departures. "The galley will load cloths and spices at Venice up to the 13th of January next," reads one government order. "She is to leave Venice on the 15th. These terms may not be extended, suspended or broken under penalty of a fine of 500 ducats." Supervised by such orders, merchantmen and sometimes naval vessels crossed the Adriatic Sea to the Dalmatian coast and sailed out of the Adriatic into the eastern Mediterranean. Wherever it gained a foothold, by treaty or force, Venice established trading posts and monopolies. Its chief stock in trade was salt from the Venetian marshes, and since, as one Sixth Century writer observed, "there never yet lived a man who does not desire salt," Venice was soon rich. The Salt Office was in fact synonymous with the state treasury.

In 1000, Venice acknowledged its ties with the sea in a ceremony that ultimately became one of the great festivals of the Venetian year. The original ceremony was one of supplication: the Doge and an entourage of officials sailed in solemn procession out into the open sea to ask God for calm, quiet waters. In its final form, however, the ceremony became a marriage, the *Sposalizio del Mare,* Venice's "wedding with the sea," in

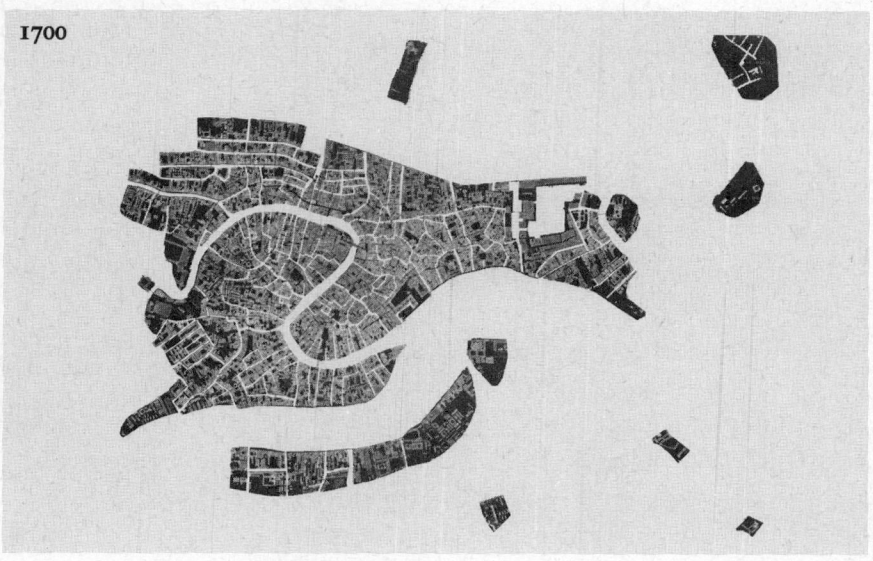

1700

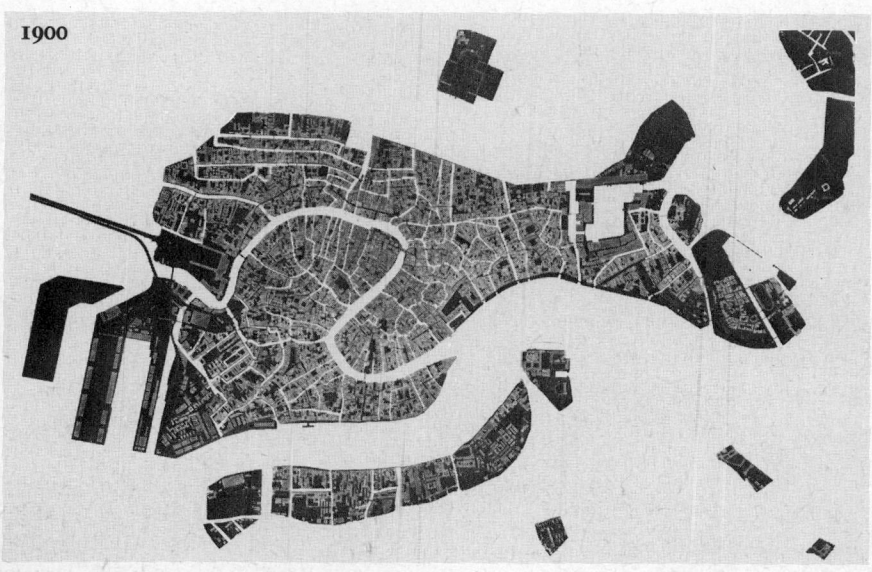

1900

which the Doge, in the scarlet-and-gold state barge, was convoyed out to the deep by all the boats in Venice and there cast a gold ring into the waves with the words, "Thus we wed thee, Oh Sea!" The first *Sposalizio* was held in 1177, when the Pope, in recognition of the city's maritime power, gave a ring to the Doge for his "beloved" and urged him to "marry her every year."

By the time of the Crusades, Venetian sea power was so great that the port of Venice became one of the chief embarkation points for soldiers headed for the Holy Land. And in 1198, when the leaders of the Fourth Crusade needed passage for many thousands of men and horses, they logically turned to Venice in their search for mass transport. Before 10,000 Venetians assembled in the Piazza San Marco, the aged Doge disclosed the Crusaders' petition. "The best people on earth have chosen us to join them in the deliverance of Our Lord," he said—and the crowd, moved as much by the occasion as the cause, cried "We consent!" But Venice did not allow its instinct for drama to outweigh its business sense. It agreed to take the Crusaders to Palestine in return for 85,000 silver marks and the promise of half of whatever territory they conquered. When the Crusaders proved unable to raise the stunning sum, Venice got them to agree to a venture more practical—and profitable—than the rescue of the Holy Land from the infidels. Joining forces, Crusaders and Venetians put down a revolt against Venetian authority in the Dalmatian port city of Zara, a prosperous Christian stronghold, and then went on to sack Constantinople, the capital of Eastern Christendom, looting it of its priceless treasures.

Through such tactics, Venice emerged from the Crusades more victorious than the Crusaders. While the Christian princes of Europe gained fleeting political control of Byzantium and the Holy Land, Venice gained long-lasting trade monopolies in most of the eastern Mediterranean; not without cause its proud historians claimed that the city controlled "a half and a quarter" of the old Roman Empire. For the next 200 years Venice fought to protect this economic empire from its chief rival, Genoa. But in 1380 a battle for Chioggia, a port very near Venice, resolved the conflict; the Genoese fleet was routed, and Venice reigned supreme. Thereafter the lagoon city turned its energies inland. As a maritime power it had come to depend almost exclusively on foreign sources of supply, an arrangement that left it particularly susceptible to blockade. Now it proposed to remedy this by acquiring territory on the mainland.

By 1420, through a series of military campaigns, political intrigues, alliances and diplomatic maneuverings, Venice managed to get control of a vast slice of northern Italy—as far north as the Alps and as far south as the river Po. Verona, Vicenza, Aquileia, Bergamo and Brescia all became Venetian satellites, and even Padua, which claimed to have founded Venice, was now simply another stone in the Venetian bastion. But the people of these cities, although they were ruled by new masters, were not unhappy with their lot; most of them preferred Venetian rule to that of the petty tyrants Venice had replaced. Venice taxed its subjects lightly and governed them with liberality—almost benevolence. In fact, when the former lords of Verona tried to recapture their domain, the people of

Verona sided with Venice to put them down. The mudbank village on the lagoon had become a world power, a state as imposing as any in Europe, a republic richer than many kingdoms.

But even as Venice grew strong, that strength was undermined. As slowly and inexorably as the foundations of its great Basilica of St. Mark sank into the ground, the city slipped into political and economic decline. Venice's expansion on the mainland involved it in a series of small but expensive wars—with the Holy Roman Emperor, the Dukes of Milan, the Marquis of Ferrara. During this same period it was also involved in a much longer and larger military involvement abroad. In 1458, when the Pope called upon Christian Europe to drive the infidel Turks out of Constantinople, Venice was the only state to respond. Its financial interests in the East left it little choice. Thus, for the next 16 years, while the rest of Europe looked on and applauded, Venice fought a lonely, heroic and losing battle. When it was over, the Venetian treasury was exhausted, and the city's trading empire in the East had dwindled to a few ports in the Levant.

More ominous still, in 1486 Bartholomeu Diaz carried the Portuguese flag around the Cape of Good Hope, opening a sea route to India. Twelve years later, sailing that route under the same flag, Vasco da Gama landed on Indian soil. The pepper and cinnamon and cloves that had once moved overland by caravan to Venetian-controlled ports and thence by Venetian ship to Europe were now accessible to anyone directly at the source. Venice faced the loss of its monopoly over Eastern trade. Predictably, "the whole city was distressed and astounded," as one Venetian diarist wrote, to learn of Vasco da Gama's feat.

However, when Titian came down from Cadore to begin his apprenticeship around 1497, Venice was still as proud and magnificent as ever, as full of spectacle, as fond of carnival. One much-traveled monk, Brother Felix of Ulm, visiting Venice after a journey to the fabled cities of the East, declared that Venice surpassed them all—was, in fact, the most beautiful city of Christendom. Other visitors, watching the splendid pageantry in the city on such festival occasions as Corpus Christi, celebrating the institution of the Eucharist, marveled at the rich garments of priests and laymen, the tunics and robes of cloth-of-gold, the brilliant colors, the masses of candles and flowers. Venetian aristocrats still gave parties and balls on the slightest pretext, from which guests emerged in glittering jewels and gorgeous clothes to dance in the streets and drift in torch-lit gondolas along the dark canals in impromptu processions of music and song.

In vain the Council of Ten, fearing for the city's financial health, forbade such extravagant displays. Venice remained pleasure-bent, an alluring, fascinating city. Its markets and shops, its quays and warehouses were piled high with all the costliest luxuries of the world—with ivory, gold and spices from the East, with silk from France and woolen goods from Flanders, with Spanish leather and Russian furs and the shining armor of Milan and Nuremberg. The Holy Roman Emperor's envoy to Venice reported that while princes elsewhere ate from pewter and earthenware dishes, nothing but gold and silver would do for Venetian gentlemen. And a French visitor was astonished to note that "when these lords want to eat, they take up their meat with silver forks."

Small wonder that Titian, stepping into this world fresh from the austere world of the mountains, preferred never to leave Venice—and when he did so for business reasons, always hastened to return. The city offered every sort of pleasure and every inducement to ambition. Even for a young apprentice there was ample opportunity for nights on the canals with the lively ladies for whom Venice was famous, lost in song and laughter and good wine. And occasionally Titian must also have looked up at the gilded palaces with their ornate façades and their gardens discreetly hidden behind high walls—and sworn to have his share.

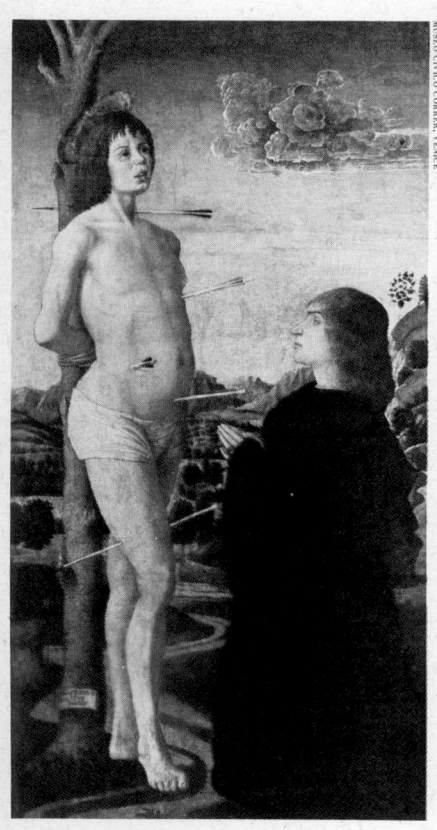

Titian's first teacher was Sebastiano Zuccato. An officer of the Venetian painter's guild, Zuccato was a respected but second-rank artist. The picture above, *Saint Sebastian Adored by a Donor,* is the only certain example of his art extant. It reveals a modest facility with anatomy, some difficulty with scale and composition and a fairly insensitive hand; his more gifted sons, however, became superb mosaicists. Titian obviously had little to learn from Zuccato, and he left his studio—perhaps at the suggestion of Zuccato himself—for the workshop of Gentile Bellini.

Titian's first master was Sebastiano Zuccato, a painter and mosaicist who was respected enough by his fellow craftsmen to have been chosen for one term as an officer in the Venetian guild of painters. Later his two sons, Francesco and Valerio, were among Titian's close friends. Like all beginners, Titian probably lived in his master's house and learned his trade the hard way. Customarily, this course of study started with such menial tasks as grinding colors, cleaning palettes, washing brushes and sweeping the workshop floor. From there, the apprentice graduated to practice in drawing—from plaster casts and life—and then to preparing the surfaces for his master's paint. Finally, he was actually allowed to paint small or routine parts of a picture himself. No painter was exempt from this training period, which ran from five to seven years. Without it, he could not be admitted to the guild, and until he was a member of the guild, he could not work. Painters were craftsmen, like potters and armorers, and the apprenticeship was the guild's way of teaching and protecting the standards of its craft.

After his apprenticeship, the painter became a journeyman and could then hire out for wages, assisting a master for another two or three years until he had progressed far enough to be a master himself. And so it must have been with Titian. At the conclusion of his apprenticeship, he would have been about 17 and, like any young man of talent, bursting with self-assurance—impatient, willful, eager to exercise his own private vision, but at the same time impressionable and ready to accept new and stimulating ideas. At this juncture—probably around 1505—he went to work in the studio of Gentile Bellini. According to the writer Lodovico Dolce, Titian began as Gentile's assistant but "could not endure following his dry and labored manner." Gentile must have shared his disaffection, for when he let him go, he remarked that Titian painted too boldly and too rapidly and would never amount to much. Gentile was then old, in his late seventies, and the most respected painter in Venice. His style was what a later age would call academic; his complex panoramas and measured processions were painted with an exactitude of line and detail that an impetuous younger artist might well have found tedious and time-consuming. Titian's own earliest work reflects these features, but it also reveals a concern for more painterly matters—in color and the way it is applied to canvas. For this reason, he was probably much more drawn to the painting style of Gentile's brother Giovanni.

At the time, the younger Bellini was still intermittently engaged in completing the series of wall paintings for the Great Council Hall, commissioned some 30 years before. The theme of this sequence of pictures had gradually evolved into an ambitious narrative of the Venetian role in the long conflict between Pope Alexander III and Frederick Barbarossa, the Holy Roman Emperor, culminating in their reconciliation at Venice in 1177—an event for which Venice claimed full credit. The sequence included the defeat of the Imperial fleet by the Venetian fleet (in a naval engagement that historians now believe never occurred); the Pope's presentation of a gold ring to the Doge, to be used in the city's "wedding with the sea"; the Emperor's submission to the Pope; the Pope's celebration of Mass in the Church of St. Mark; and the triumphant arrival of the Pope, the Emperor and the Doge at the gates of Rome with, in Vasari's words, "eight standards of various colors and eight silver trumpets . . . a large number of horses and a vast body of soldiers"—another event that had taken place only in the lively Venetian imagination. Unfortunately, all these paintings were destroyed by a great fire in 1577; contemporary accounts describe them as marvels of beauty.

They were not, however, the only sort of work Giovanni was doing. He was also painting a distinctive type of Madonna, round-chinned and innocent, looking at her plump Infant from under lowered lids, altogether appealing *(page 23)*. Sometimes the Madonnas are set against plain backgrounds, sometimes against a curtain with a glimpse of landscape beyond it. Their clothing is simple; a veil covers their hair and over the veil is thrown a mantle, rather like a shawl—except that the richness of its color bespeaks a certain elegance of texture. Once in a while there is a residue of Byzantine influence in the stylized angle of the head or the shape of the hands, but it is a Byzantine influence that has been transformed and humanized.

In addition to the Madonnas, Giovanni was painting equally human portraits of prominent Venetians—aristocrats in black robes and caps, a heavy-jowled professional soldier, the aged Doge Loredano with a slight smile on his wise, bony, wrinkled face *(page 24)*. Giovanni's portraits were so widely admired that, according to Vasari, there was scarcely a wealthy home in Venice without one; Vasari even credits the artist with introducing Venice to the custom of memorializing the most consequential members of famous families in paint. True or not, Giovanni must certainly have had trouble keeping up with his commissions, for beside portraits, Madonnas and scenes of Venetian history, his services were also in demand for altarpieces. It is a measure of his importance that one of the most knowledgeable and avid art collectors of the day angled for his work ferociously; the patronage of Isabella d'Este, Marchioness of Mantua, a small city-state 80 miles from Venice, has in fact provided a strong clue to the link between Giovanni Bellini and Titian.

Isabella and her husband Francesco Gonzaga made their court one of the great centers of Italian culture. She and the marquis were both people of taste and perception, and both came from families with a long tradition of patronage of the arts; both were also astute and able rulers. Isabella herself was one of those fascinating women who attract creative and intellectual men—like Eleanor of Aquitaine, long before and

Madame de Staël, long afterwards. Learned, witty, generous but strong-willed, Isabella dealt equally effectively with ruthless princes like Cesare Borgia and gentle artists like Raphael. Her friends included many of the most brilliant men of her day, and her palace was a showcase for the work of the best painters in Italy. She could be high-handed and arrogant, but on the other hand her judgment of art was exquisite. The many appreciative letters and poems addressed to her by artists indicate an awareness of the value of her sponsorship.

Isabella began trying to get a painting from Giovanni in 1496. He made vague promises but did not actually agree to accept a commission until 1501—and even then he turned down Isabella's suggestion for a subject, an allegory drawn from classical literature. She offered to let him pick his own theme, provided it came from the "antique" and had a "fine meaning," but when this proposal produced nothing she finally settled for a Nativity. In 1505, undaunted by her previous failure, she tried again for her allegory. This time she used two poet-emissaries—Pietro Bembo and Paolo Zoppo. After a short interval Bembo was able to report that "the fortress would soon surrender" if Isabella would write Giovanni in her own hand. She did, but again negotiations were protracted. Isabella insisted on something allegorical; Giovanni protested that he was a painter of religious works, with no feeling for pagan themes. "The treatment of the theme . . . will be dictated by the imagination of the man who is painting the picture," he told Bembo to write her. "He does not like having his style cramped . . . being used to taking his own line in his painting."

Some historians believe that Giovanni did at last give in and begin to work on the allegory. Although there is no record of Isabella's ever having received such a painting, a picture in the castle of her brother Alfonso d'Este, Duke of Ferrara, may have been the one Isabella ordered. It is called *The Feast of the Gods* and is unlike anything else Giovanni ever painted. Its theme is drawn from Greek mythology, and it shows a group of gods, goddesses and nymphs dining *al fresco*. The painting was received and paid for by the Duke in 1514, and sometime after that sections of it were altered and repainted by Titian—for Giovanni, in the meantime, had died. Because of this known link between the two artists, some experts speculate that Titian may also have worked on the original commission as Giovanni's assistant and that he may have had a hand in other paintings by Giovanni as well.

At the same time, the young journeyman was also accepting commissions of his own. One of the earliest was a painting done for Jacopo Pesaro, Bishop of Paphos, a leading member of one of the great Venetian families. The picture shows the bishop being presented to St. Peter by the infamous Borgia Pope, Alexander VI—and curiously it says as much about the artist as it does about his subject. Titian has placed St. Peter on a dais decorated with a classical frieze in deference to the classical mood of the times, and he has portrayed the Pope as a benign ecclesiastic, although Alexander's mistresses and political intrigues had made him hated and feared throughout Italy. Also, Titian seems to have changed his mind about style halfway through, for the figure of St. Peter, on the left, is tight and pinched while those of the bishop and the Pope, on the right, are done in a much freer and more flowing line.

How long Titian worked in Giovanni Bellini's studio is uncertain. He may still have been there in February 1507 when Gentile Bellini died, bequeathing to his brother—among other things—the task of completing his last work, a painting for the Scuola of St. Mark. Giovanni obeyed his brother's request, but Gentile's real artistic heir was another prominent painter. Vittore Carpaccio was a master of the great crowd-filled wall paintings that had been Gentile's hallmark. But Carpaccio, in addition to being a superb reporter, was a remarkably inventive painter. His paintings, regardless of what they ostensibly illustrate, are really pictures of Venice. In their color and richness of detail they record the physical appearance of the city as it actually was. They also record a side of the Venetian character for which Titian was to provide the counterpoint. Carpaccio was concerned with the outward trappings of reality, with the show of wealth and the practical aspects of business and religion, just as Titian somewhat later was concerned with color and human emotions and the pleasures of the senses.

Carpaccio was a native of Venice, born into a family of fishermen and boat-builders. His early life, like that of most artists of his day, is shrouded in mystery. His first major commission was a series of eight large canvases for the Scuola of St. Ursula, a fraternity of merchants and craftsmen whose meeting hall was adjacent to a Dominican church. In 1488—the presumed year of Titian's birth—the Scuola decided to decorate its hall with a series of paintings depicting the history of its patron saint, and hired Carpaccio for the job. The story of St. Ursula, much

Giovanni Bellini brought a unique quality to the painting of mythology. His *Feast of the Gods* (*left*) is peopled with down-to-earth human types rather than the idealized heroes of Raphael, Botticelli and Mantegna. Indeed, some of Bellini's celebrants here have been identified as his contemporaries: the god Silenus standing behind the donkey has been recognized as Pietro Bembo; the goddess at center holding a quince may have been Lucrezia Borgia, and the figure seated behind helmeted Mercury has been seen as the artist himself. Whether or not this guessing game proves true, it is evidence that Bellini added warmth and reality to classical subjects.

embroided by legend—and by Carpaccio—tells of a young Breton princess who has been betrothed to the pagan king of England on condition that the king be baptized a Christian and that she be allowed to make a pilgrimage to Rome. Accompanied by an entourage of 10 virgins, each of whom takes with her 1,000 serving maids, Ursula sets out on a journey that takes this chaste army down the Rhine. At Cologne a storm forces them to land, and Ursula is visited by an angel who orders her to return to Cologne for martyrdom after she has completed her pilgrimage. She obeys and is met on her return by Attila and his Huns, whose westward march had brought them to the Rhine. Ursula's fate is held briefly in check when Attila's son falls madly in love with her; but she repulses him, and she and her virgins die under the Hunnish swords.

Carpaccio's illustrations for this tale are a splendid panorama, not of some faraway land, but of Venice *(pages 20-21)*. The people who flocked into the hall of the Scuola to view the paintings when they were hung in 1498 must have felt as though they were looking through huge windows upon dazzling perspectives of their own city. Venice is in every picture. When the envoys from the English king present themselves at the court of Ursula's father in Brittany, there are gondolas in the water behind them; when they return to England, it is an England full of Venetian architecture and shipping, Venetian canals and bridges. Among the crowds of rudely dressed Britons are young noblemen dressed in silks and velvets and wearing the pearl-encrusted insignia of Venice's aristocratic clubs, the *Compagni della Calza* (literally Companions of the Stocking, on which the insignia were customarily affixed). It has even been suggested that the figure of Attila's son in the scene of Ursula's funeral is actually a self-portrait of Carpaccio.

Venice was delighted with these pageantlike pictures of familiar sights —dogs and pet monkeys, lounging cavaliers, water and boats and flying flags—and Carpaccio's future was assured. For the Scuola of the Slavonians, he painted a history of St. Jerome; for the Scuola of St. Stephen, a history of St. Stephen; for the Scuola of the Albanians, scenes from the life of the Virgin. In 1494 the Scuola of St. John the Baptist commissioned him to paint a picture of one of the miraculous happenings involving its dearest possession, a relic of the True Cross. The central event of the painting shows the Patriarch of Grado, on his balcony overlooking the Grand Canal, using the relic to exorcise the devil from the soul of a tormented man. But the viewer of Carpaccio's opus tends to let this event go almost unnoticed in favor of the scene the artist has painted below the balcony. Passers-by go about their business, cavaliers lounge in idle talk and gondoliers pole their customers through a traffic-filled canal. The Rialto Bridge bustles with people, hinting at the shops out of view, and a swinging tavern sign conjures up good food and drink. In one of the gondolas the inevitable lap dog curls up cozily, and above the canal, suspended from long poles, household linens bleach in the sun. The skyline bristles with chimney pots.

This is Venice as it really was—practical and sensual, poetic and prudent—a city confirmed in its belief that miracles were all very well, but that a man must get on with the business of making a living. There are

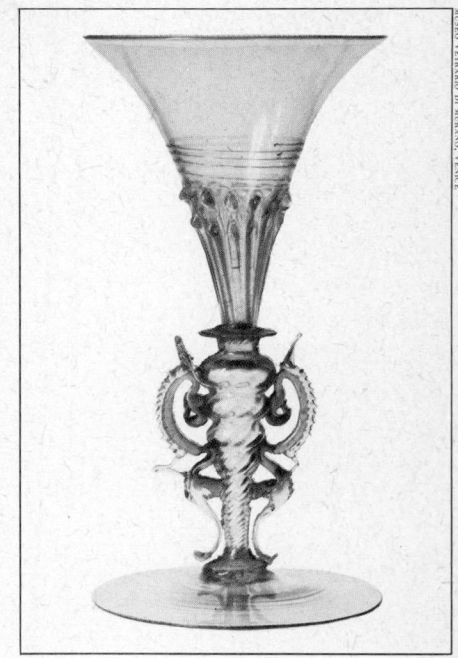

many stories of this aspect of the Venetian character. One tells of a Venetian fisherman who ferried three saints across the lagoon, during which trip his passengers destroyed a whole boatload of devils. It was a fine performance, the fisherman said when they reached the other shore, but which saint was going to pay the fare? Similarly, there is a story of a Venetian who visited Florence during the days when Savonarola was preaching the sinfulness of worldly possessions. As he watched wealthy Florentines throw their books, paintings and fine clothes on huge bonfires, the Venetian remarked that he would be happy to take over these articles if their owners no longer wanted them. The Florentines accused him of being frivolous about religion, but, in fact, he was just being Venetian. He was displaying that respect for the good things of life that tempered Venetian piety. *"Veneziani, poi Cristiani,"* said Venice: "We are Venetians first, then Christians." This was the shrewd, sensible Venice memorialized by the paintings of Carpaccio.

Venetian glass, prized for centuries by collectors, is made on the island of Murano (*map, pages 38-39*) with skills that were ancient even when the glassmakers' guilds were formed in the 13th Century. Murano became the home of the industry because Venetians on the main islands feared the possibility of city-wide fire from the glass furnaces. Moreover, Murano's semi-isolation guarded the intricate secrets of glassmaking. In Titian's day Venice introduced *cristallo,* the first near-colorless glass, which could be blown to unprecedented thinness and worked into a myriad of forms. At the left, above, is a 16th Century oil lamp in the shape of a horse; a wick was probably inserted in the tail. The delicate, clear glass chalice with its ornate stem may have served as a vase.

If it had been the only Venice, Titian's career might have been different. In the service of men who were shrewd and sensible, he might have trimmed his sails to paint in the manner of Carpaccio. Fortunately for Titian, whose talents lay elsewhere, there was another side to the Venetian character, a side responsive to poetry, to pastoral quiet, to the more introspective pleasures of life. This was the side reflected in the gentle, dreamy Madonnas of Giovanni Bellini, and even more so in the idyllic landscapes of Giovanni's most precocious student, Giorgione of Castelfranco. Giorgione's luminous pictures captivated Venice and awakened contemporary artists, Titian most notably among them, to a new painting style. Under Giorgione's influence, Titian abandoned any attempt to emulate the meticulous paintings of men like Carpaccio and began to realize his own artistic powers.

Venice Observed

Titian's Venice, like a luxuriant, parti-colored blossom sprawled on a jade-green lily pad, lies fragilely moored between sea and land on an Adriatic lagoon. For over 1,400 years she has weathered tides, storms, wars and the rising and ebbing fortunes of her own merchant-based economy. For over 1,300 of these long years she remained a republic, the longest lasting in the history of man's government. Half-Oriental, half-Western, Venice sits on the very crossroad of two worlds. Her art, her architecture, her institutions and the character of her people reflect this tradition-rich dual heritage in a way that is rare among cities.

Founded by desperate handfuls of Northern peoples fleeing to the safety of offshore islands from the devastations of mainland invaders, the city evolved strong traditions of independence and personal liberties. Elected representatives, whose powers were limited by a vigorous system of checks and balances, formulated her policies. Enterprising citizens, even those fairly low on the social scale, could make their way in trade, fishing, shipbuilding and the salt and glass industries. The two vast Venetian fleets—one for war and one for commerce—kept the island republic safe from invasion and filled her coffers with gold and silver. During Titian's time and for a few centuries afterward Venice's unprecedented wealth led to a way of life that was secure, elegant, gaudy and filled with almost daily ceremony and spectacle.

Distinguished by a row of bronze griffins above the Venetian Gothic stonework of its second-story windows, the Cavalli-Franchetti Palace presents a bold front to the Grand Canal. The gaily striped pilings in front of this elegant residence are used to moor gondolas, the traditional Venetian substitute for taxis. In the background, marking the entrance to the Grand Canal, are the two domes and bell towers of the Church of Santa Maria della Salute.

Legend has it that the early settlers in Venice, an uprooted people called the Veneti, who were trying to escape pursuing raiders, were told by a supernatural voice to climb a tower on the mainland from which they would see a place of safety. Sure enough, on climbing a tower that was miraculously there, they spied some green and brown patches in the lagoon, which they recognized as islands. Piling into boats, which also appeared providentially, they rowed to salvation. Their first settlement, called Torcello, or little tower, can be seen in the upper right-hand corner of the map below.

In subsequent years the hardy band of pioneers mingled with an indigenous tribe of primitive fisherfolk and spread out to other islands. Eventually they concentrated on the two largest ones, which constitute Venice proper.

Divided by the serpentine Grand Canal, Venice is sliced

through by some 170 smaller canals. The spiritual and political heart of the city is the Piazza San Marco—seen at the center of the map and in an aerial photograph on the following pages. In the Piazza are the Basilica of St. Mark, with its soaring bell tower; the palace of the Doge, Venice's head of state; and the colonnaded administrative buildings of the city government.

Other landmarks that were especially important in Titian's day were the Arsenale—the walled enclosure seen on the upper edge of the island at far right—where thousands of Venetian galleys were built, armed and sent to sea; the salt warehouses—on the tip of the island obliquely facing the Doge's Palace—whose revenues provided a stable base for Venetian trade; and the angular Rialto Bridge, which spanned the Grand Canal and connected the two parts of the city's principal business district.

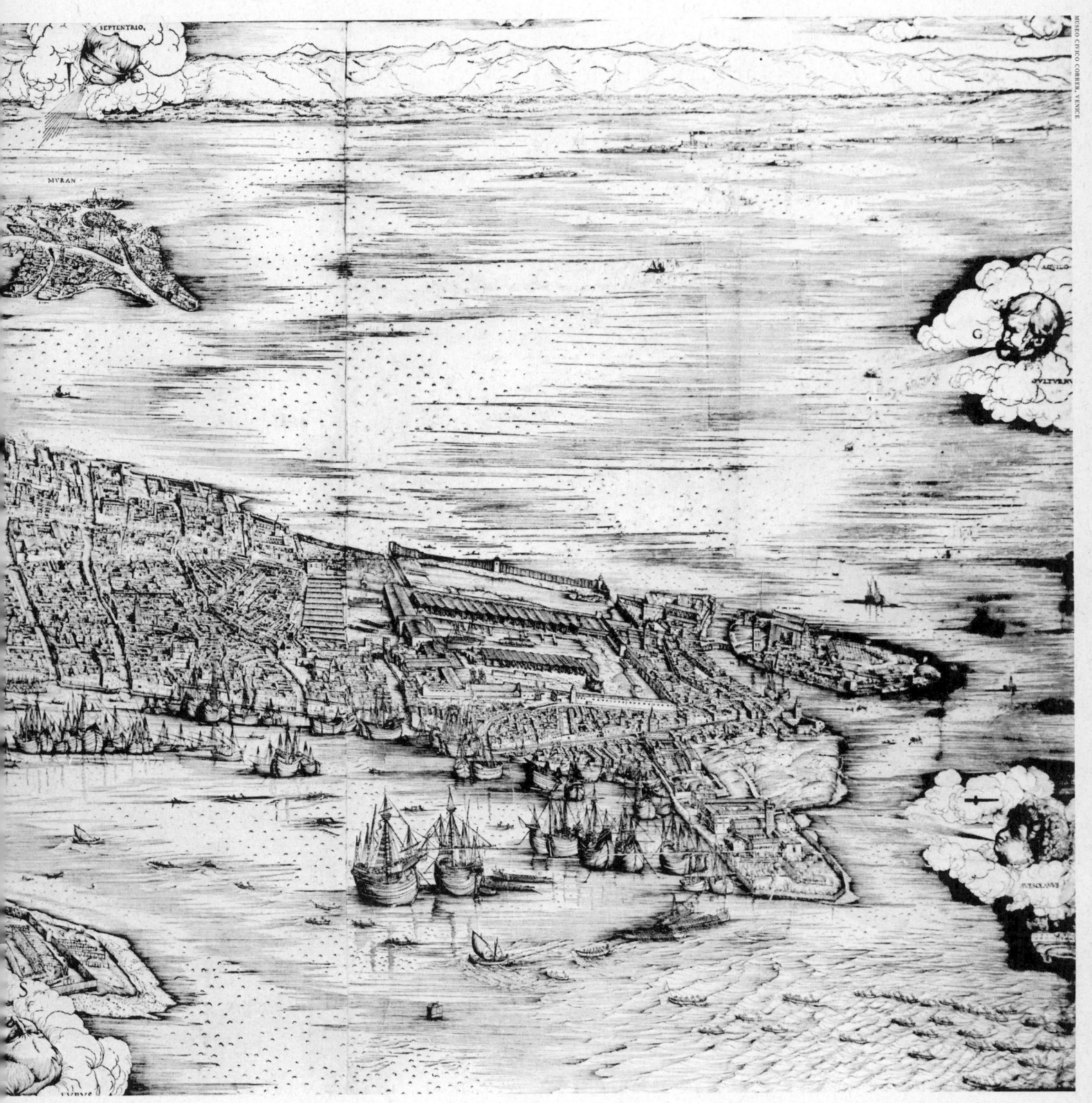

Jacopo de' Barbari: Map of Venice, c. 1500

39

Unlike the Basilica of St. Mark, which was designed and executed to a single architectural plan, the Doge's Palace is a conglomerate of different styles. This is at least partly due to the fact that the palace was rebuilt over the course of many centuries, both in accordance with the growing administrative needs of the city, and also because accidents destroyed parts of the buildings. As early as 820 A.D. there was a fortified chapel on the site. In almost every succeeding century new additions were made, most of them destroyed in 1419 by fire. It was then that the basic Gothic character of the palace was mixed with a Renaissance classicism; when pointed Gothic arches were supported by regular classical columns, lacy open stonework was mixed with pink and pearl-colored marbles in intricate foliated patterns. This eclectic character is illustrated in the view *(above)* of the inner courtyard. During Titian's own lifetime the building was severely damaged by fire, and several among a committee of

architects proposed that the palace be rebuilt in a "modern" style. Fortunately, the proposal was vetoed and the building kept its original character. As the 19th Century French historian Hippolyte Taine wrote: "There is nothing here that is bare and cold, everything is covered with statues and reliefs; the pedantry of learning or of criticism not having intervened . . . to restrain the fire of the imagination or the desire of giving pleasure to the eye. Venice . . . wished architecture to possess and delight every faculty, and decked it with ornament, column and statue, made it a thing of riches and joy."

Elegance and richness are especially evident in the staircase at the left. The main entrance to the palace from the courtyard, this beautifully proportioned stair, called the Giant's Staircase, was where the Doge was crowned. Here, flanked by statues of Mars *(left)* and Neptune *(right)* representing armed power and control of the sea, the Doge swore to lead the city fairly and honorably.

The spirit of independence that had characterized Venice since the time of her founding was sorely tested during the centuries before Titian's day. As several families, by enterprise or good fortune, gained extraordinary wealth, they also sought to control power. And although the tradition of electing a head of state—the Doge, or duke—was an old one, in the course of the 13th Century it became clear that in the future the city could be seriously divided by brawling between the partisans of the various great families and that corruption of the voting might occur. To suppress these rivalries, Venetians were forbidden to display in public the coat of arms or badge of any family. Furthermore, and more important, in 1268 an elaborate system of voting for the Doge was devised that made it impossible for any one faction to influence the election significantly. The system is diagrammed below. First, the members of the city's Great Council, composed of representatives of the important families and other worthy citizens, select 30 names by lot. Those 30 select 9,

also by lot. It is clear that even at this early stage it would be beyond any individual's power to "rig" the election. But the Venetians were cautious; and so, the 9 vote for 40, the 40 are reduced by lot to 12, the 12 vote for 25, and so on. Finally, 41 voters elect the Doge, who must receive more than 25 votes to win.

After the election was completed, the Doge was carried ceremonially into the Basilica of St. Mark where he was introduced to the people, heard Mass and swore his oath. He was then carried around the Piazza and through the ornamental archway at the left—which is called the Porta della Carta—into the courtyard of the Doge's Palace, where, on the Giant's Staircase, he was crowned.

Despite all this pomp and circumstance the Doge was politically restricted in his powers. He was dressed resplendently by the people of Venice—who revered him extravagantly—and served at all their public ceremonies, but, ironically, in his oath of office he virtually swore never to interfere in the city's government.

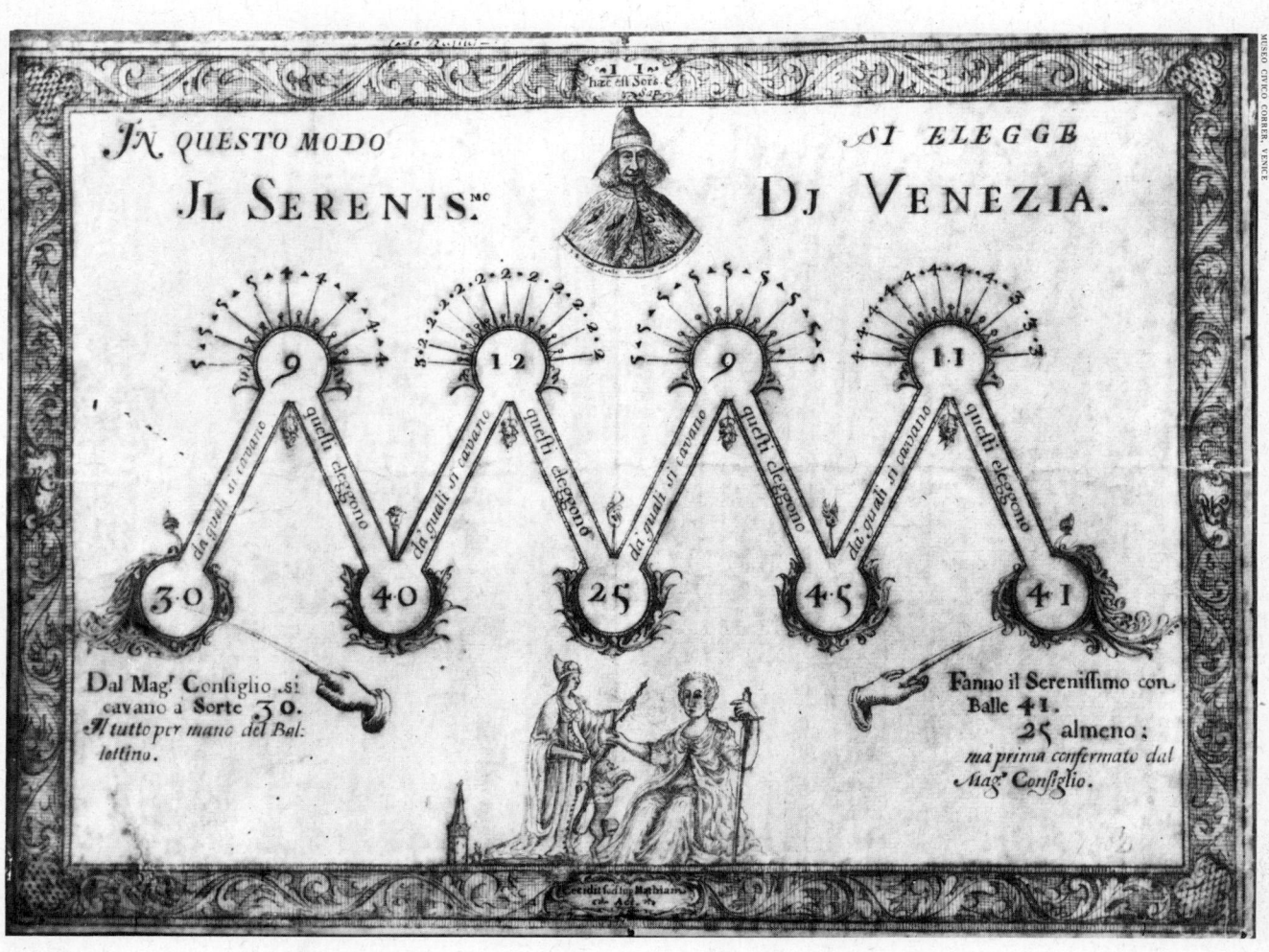

Spectacle, ceremony and sport were vivid facets of Venetian life in Titian's day. Taking place on the canals, bridges and open plazas—called *campi,* or fields—religious celebrations, traditional festivals and active games thrilled the populace. Here and on the following pages are contemporary engravings of some of these activities. At the left, below, is a brutal game in which the adherents of two factions in the city—the Castellani and the Nicolotti—

PAGES 46-49: FROM A VOLUME OF ENGRAVINGS BY GIACOMO FRANCO, PUBLISHED IN THE EARLY 1600s. MUSEO CIVICO CORRER, VENICE

fought for possession of a bridge. Armed with sharpened sticks the combatants sought not to kill one another but to drive the opposition off—a kind of dangerous king-of-the-mountain game. At the right are a host of sports events—grabbing a hanging goose *(foreground)*, bull-running *(background)*, bearbaiting *(left center)*—as well as group dancing *(center)* and other popular competitions. Spectators crowd the overlooking windows and balconies.

Games and entertainments figured prominently among the outdoor activities of Venice. At the left, below, is an illustration of a kind of football game (the player near the right-hand archway has the ball), which was played by an exclusive group (contestants were required to be of noble birth) during Lent in the courtyard of the Church of Sant' Alvise. At the right is a composite view of some of the performances that were available to the public as they

strolled through the Piazza San Marco. The engraver noted in a caption that Venice is "the only nation that does not object to such performances morning and night." He also identified with captions the nationalities of some of the onlookers—Greek, French, Spanish, Turkish, English and a Capelletto, a local militiaman. They are watching musicians, a snake charmer, a magician, actors and *(at center)* a trained dog act.

III

A Fresh Eye
on the World

One of few drawings by Titian that still exist, this is a pen and ink study that the artist made for a fresco, which he was commissioned to paint in the Scuola del Santo in Padua. Not only in its active subject—a furious husband stabbing his adulterous wife—but also in the vigor of its line, the drawing reveals Titian's early interest in physical energy and psychological drama.

Study for *The Jealous Husband*, c.1511

After his stint in the studios of Gentile and Giovanni Bellini, Titian moved on, around 1507, to become assistant to one of the most controversial and enigmatic figures in the whole history of Italian art. Giorgione of Castelfranco has been credited with many paintings, but only a few of them have survived, and of these only one, *The Tempest,* can be documented as a genuine Giorgione. As for the rest, either their authenticity has been questioned, or they have been shown to be retouched or repainted by other hands. Giorgione, who never signed or titled his paintings, has also puzzled scholars by failing to make their meanings explicit. Some of his works have gone by as many as six different and contradictory names—and occasionally a scholar claims to see in a Giorgione painting things that are not really there. The information about the artist's life is equally confusing. He has been described as a peasant's son and a nobleman's bastard, a licentious man and a chaste one, a victim ultimately of the plague and of jealous grief over the loss of a mistress. One critic has even predicted that someone will someday argue that Giorgione was not a man but a myth.

No one, however, denies the importance of Giorgione's contribution to art. From his own time to the present, scholars have generally agreed that he profoundly affected the work of other painters and indeed the entire course of painting. He has been called an innovator, a poet, a harbinger of romanticism and the first painter to paint the human spirit. "All the more eminent artists confess," writes Vasari, "that he copied the freshness of the human form more exactly than any other painter, not of Venice only, but of all other places." The modern historian George M. Richter has acclaimed Giorgione's art as the quintessence of Renaissance humanism.

Giorgio (the name Giorgione, "Big George," came later, with fame) was born around 1477 or 1478 in the town of Castelfranco, not far from Venice. Despite romantic notions of his noble origin, his parents were apparently common people. At some point in his youth he came to Venice, studied painting, and was much influenced by the style of Giovanni Bellini. He worked for a time as a decorator in the furniture shops in the

Rialto, the city's business district, painting scenes on cupboards and chests. He also seems to have won notice for his frescoes, using the façade of his own house as a showcase for murals and "ovals containing musicians and poets, and also fantasies in chiaroscuro." Vasari reports that he was pleasing in manner, fond of the ladies and a gifted singer who accompanied himself on the lute. Prominent Venetians eagerly invited him to their homes.

Giorgione was very early attracted to the new technique of painting in oils. He used it in several paintings of the Virgin and in a head of David for which he may have served as his own model. According to Vasari, he also inserted an oil painting into the fresco for the outside of the sumptuous private palace of the Soranzo family; this addition, an unorthodox decoration for an exterior, "weathered quite well," Vasari notes. Giorgione's work had a distinctive quality that quickly gained him a following. It was poetic and still, full of golden light and deep shadows, rather like a reverie brought to life. Giorgione's colors were rich and warm, and his forms blended into the atmosphere as smoothly as the forms of Leonardo da Vinci. In fact, the two great painters may have compared notes on their techniques when the great Florentine artist visited the lagoon city in 1500.

The demand for Giorgione's work soon grew so that he had to hire assistants. Among them were two young men who fell deeply under his spell and whose work became almost indistinguishable from his. One was Sebastiano del Piombo, who later moved to Rome and found favor with Pope Clement VII; the other was Titian. When Giorgione died in his early thirties, probably of plague and, as Vasari put it, "greatly to the loss of the world, thus prematurely deprived of his talent," he left behind for consolation these "two worthy disciples and excellent masters, Sebastiano . . . and Titian of Cadore, who not only equaled but surpassed him."

These two unidentified figures were copied from Giorgione's frescoes on the Fondaco dei Tedeschi before the originals finally weathered away. Published as etchings in 1760 by Antonio Zanetti, they are no more than faint echoes of Giorgione's brilliantly colored murals.

Titian's association with Giorgione probably began when Giorgione was commissioned by the Venetian government to decorate the façade of the newly built Fondaco dei Tedeschi. Although the name of this building literally meant "Warehouse of the Germans," the Fondaco was actually a hostelry, a home away from home for the German merchants in Venice; they slept there, ate there, quartered their servants and horses there, used its storerooms for their merchandise and its public rooms as places of business. In fact, the city did not permit its German population to live or work anywhere else. This regulation, in effect since the 13th Century, was passed partly for the Germans' own protection, but in equal degree for the safety of the Venetian state. The suspicious Venetians did not trust foreigners; most especially they did not trust Germans who more than once had crossed the Alps in military force to establish footholds on the Italian peninsula. The Fondaco permitted the city to keep the business transactions of these aliens—as well as their social lives— under constant surveillance. Indeed, the staff of the Fondaco, from its supervisors to its kitchen help and stable boys, was a highly efficient espionage system for the Venetian government.

In 1505 the building that had long served as the Fondaco burned to the ground, and the Venetian government quickly authorized construction of a new one. It allocated a space for it on the Grand Canal and retained a German architect, Girolamo Todesco, to draw up plans. Girolamo designed an impressive rectangular building with an open court and many rooms. Its lower floor contained storage facilities and servants quarters; its upper floors contained two handsomely proportioned public rooms and 80 bedrooms, which rented for prices ranging from eight to 10 ducats a year—the higher the floor, the less the rent. The Venetian government stipulated, however, that the building was not to be decorated with any marble or carving or fretwork—possibly to keep down the cost, possibly to prevent it from outshining in splendor the neighboring *palazzi*. Giorgione's commission to decorate the Fondaco with murals resulted from this decree.

Giorgione hired Titian to assist him with the Fondaco decorations, and the project, begun in 1507, was apparently completed the following year. Nothing is left of it now but a single fragment, the figure of a nude female, which hangs in the Venetian Accademia. But when the murals were fresh and new, Venice admired them extravagantly. In 1658 a Venetian engraver, Giacomo Piccini, copied two of the figures frescoed by Titian, and about a century after that, all the murals that remained were memorialized in a series of etchings by a nobleman and scholar named Antonio Zanetti. The work of both artists, Zanetti wrote, was remarkable, but "whilst Giorgione showed a fervid and original spirit and opened up a new path over which he shed a light that was to guide posterity, Titian exhibited in his creations a grander and more equable genius which placed him in advance of his rival, on an eminence which no later craftsman was able to climb." The praise heaped on Titian's part of the Fondaco murals was, in fact, said to have made Giorgione jealous. Lodovico Dolce reported that Giorgione sulked in his room for days. But this seems unlikely behavior for a man of Giorgione's cheerful temperament—and besides, it is a story often told about masters and their gifted assistants.

In 1541, when Vasari visited Venice, he had other, somewhat less complimentary things to say about the Fondaco murals. Although he conceded that they were "well designed and colored with great animation," he accused Giorgione of thinking only of "executing fanciful figures, which would show his ability. There is in this work, indeed, no arrangement of events or even single episodes of Venetian history. I, for one, have never known what his pictures mean, and no one has ever been able to explain them to me." Vasari's complaint touches upon one of the most intriguing aspects of Giorgione's work. Many of his paintings do indeed seem to be expressions of the artist's own vision rather than some external event. Their composition and content seem dictated by subjective concerns—the immediate joys of stroking color on canvas, of forming attractive designs and shapes, of exploring the rich contrast between light and shadow. To modern audiences, conditioned to paintings with titles like *Black on Black* or *Opus #4,* this approach seems entirely valid. But the audiences of Giorgione's time—and of many generations thereafter—expected paintings to be pictures of people and subjects they could identify.

In their search for the themes missing from Giorgione's paintings, scholars have proposed themes of their own. Some have been factual, some romantic, and some have occasionally been misleading and contradictory. In a few cases scholars have even betrayed a certain confusion about what their eyes actually saw. Thus the painting commonly known as *The Three Philosophers (page 67)* has sometimes been called *The Three Wise Men,* which it very well may be. But it has also been called *Three Astronomers Observing an Eclipse,* although a round sun is clearly visible, setting on the far horizon. Similarly, *The Tempest (page 66)* has been said to portray a soldier and a gypsy, although the man in the painting is neither armed nor in uniform, and the woman is nude except for a shawl. As for the famous *Fête Champêtre (pages 72-73),* a painting in which Titian had a hand, it has been described as a country picnic, an impromptu open-air concert, a meeting of poets and their muse, and a casual dalliance between two rosy country girls and two boys from town.

These strange, lovely scenes, timeless and golden as daydreams, are Giorgione's most important contribution to art. They are pictures to be looked at for their visual satisfactions alone, and as such they suggest that art need have no other purpose. Giorgione's figures are neither actors nor symbols; they are simply elements in a landscape; his nudes, for all their melting flesh tones, are not objects of desire but perfect forms, beautiful people in a beautiful world. For the cultured men and women of the Renaissance, these pictures must have provided the same refined pleasures as poetry and must have summed up all the qualities they held in highest esteem. In the words of the critic Bernard Berenson, Giorgione's lyrical paintings were "perfectly in touch with the ripened spirit of the Renaissance" and succeeded as only those things can succeed that both awaken and satisfy a need.

Yet Giorgione was not just a painter of serene dreams. Like his fellow artists he accepted commissions for portraits and altarpieces *(page 65).* In the religious paintings, the sense of the spiritual is so strong that the figure of Jesus in an altarpiece for the Church of San Rocco, for instance, was widely believed to work miracles. Nor was this the end of Giorgione's skill. He once produced a painting to settle a philosophical argument. When some sculptors he knew claimed that their art was superior to painting because it permitted the viewer to see more than one aspect of the body, Giorgione, in rebuttal, painted a picture of a nude—long since lost—in which four aspects were visible at once; while the figure had its back to the viewer, the front was reflected in a stream, the left side was reflected in a piece of shining armor and the right side in a mirror.

Titian may or may not have continued to work with Giorgione after the Fondaco murals were finished, but he certainly continued to be influenced by Giorgione's style. Giorgione, in fact, provided the key that unlocked Titian's talent. Just as, earlier, Titian had turned away from the prevalent painting style represented by Gentile Bellini—a style that was stiff, dry, precise—so he now turned away from the prevalent practice of beginning each painting with a series of carefully detailed preliminary studies. Giorgione preferred to rough out his compositions in color directly on his canvas. He believed, says Vasari, that "to paint with colors only, without drawing on paper, was most perfectly in accord with the true principles of art." To Titian, this approach was irresistible. For a time his work was so like Giorgione's that it was difficult to tell them apart. Vasari notes of one particular portrait of an elegantly dressed man *(page 113)*—once thought to be of a member of the Barbarigo family, patrons of Titian—that the work "would have been taken for a picture by Giorgione if Titian had not written his name on the dark ground."

In Titian's hands, however, Giorgione's style underwent two subtle changes. Where Giorgione chose to leave his meanings obscure, Titian preferred to make his literary references explicit. And where Giorgione's female nudes are idealized, Titian's are robustly human. For example, when Giorgione died he left behind an unfinished painting completed by Titian, the *Sleeping Venus*. As painted by Giorgione the goddess of love was almost an abstraction, delicately drawn, smoothly contoured. Without changing the figure, Titian emphasized its eroticism by adding in sumptuous silken draperies on which the recumbent Venus lies. To Titian, a painting was clearly a commodity as well as a work of art.

In 1510, the year Giorgione died, Titian left Venice for a time to seek his fortune elsewhere. Probably he was restless, like all young men. Perhaps he thought to get bigger jobs in smaller towns where the competition was less keen, and so to improve his bargaining position at home. He might have picked a better time; conditions on the road were not entirely safe. Venice had just come through one desperate war and was facing another. The city's expansion on the mainland had brought it the inevitable harvest of success: the rest of Europe had become alarmed. In 1508, in the French city of Cambrai, a league had been formed to curb what its members called "the insatiable cupidity of the Venetians and their thirst for power." Having done so, they proposed with equal cupidity to divide up Venice's territorial holdings among themselves. The most active members of the league were Pope Julius II (justly famed as the "warrior-Pope"), Louis XII of France and the Holy Roman Emperor Maximilian; other co-signers included the Kings of Spain and Hungary, and the Dukes of Ferrara and Savoy.

The Pope had opened hostilities against Venice in 1509 by excommunicating the city, but Venice refused to recognize the papal decree and set guards to prevent it from being posted. Next, France sent an army across the Adda River, Venice's western boundary. The invaders engaged the Venetians at the battle of Agnadello, 150 miles west of Venice, and completely routed them. By June 1509 papal armies had taken over the formerly Venetian-held cities of Ravenna, Faenza and Rimini; Louis XII had annexed Brescia, Bergamo and Cremona; and Padua, Vicenza and Verona had become part of Maximilian's Holy Roman Empire. Only the natural protection of its lagoon saved Venice itself. But then the redoubtable Julius II suddenly made peace with Venice. Having got what he wanted out of the League of Cambrai, the Pope now became aware of the danger of the presence of a French army on Italian soil. Maximilian soon aligned himself with the Pope against the French, and so did the

King of Spain—so that the old League of Cambrai was in effect turned inside out.

All through this tumultuous period, life in Venice went on very much as usual. The wealthy danced and supped in an endless round of entertainments. The city's treasury might be depleted and its very life at stake, but its citizens appeared not to notice. Perhaps Titian was equally heedless of the personal dangers of traveling at so perilous a time; perhaps wars and politics lay outside his field of interest. In any case, the city he chose as his destination when he set out from home in 1510 was a city newly returned to the Venetian fold and newly at peace: Padua.

In Padua, Titian obtained commissions for several frescoes, taking as an assistant a local Paduan artist, Domenico Campagnola. A memorandum on the back of one Titian sketch notes that Domenico had been paid a debt owed him by Titian "for assistance in working on the front of the Cornaro Palace in Padua," and a few lines on the back of one of Domenico's drawings indicated that he worked with Titian in 1511 on frescoes for the Scuola del Carmine. Their association may have extended to other work as well, for Domenico's later paintings are so imitative of Titian's that they were often mistaken for his work. Titian's most important project in Padua, however, was a trio of frescoes for the Scuola del Santo.

Water dripping through the roof of the Scuola has irreparably damaged these paintings, but they are still remarkable. In their colors and composition they resemble Giorgione's work, but Titian's own emerging

Reproduced below and greatly reduced in size—the original is over 100 inches long—is the right half of Titian's monumental woodcut, *The Triumph of Faith*. The crowded, active procession is led by Adam and Eve. They are followed by Noah (holding the Ark), Moses (brandishing the tablets of the Ten Commandments), Abraham (wielding a sword), a group of pagan sibyls, learned prophets and innocent children accompanied by a host of trumpeting angels. At the left, preceding the chariot that bears Christ, is a muscular youth, probably the Good Thief, carrying a heavy cross. Titian undoubtedly intended his frescolike woodcut to rival the familiar triumphal scenes that apotheosized pagan leaders such as Julius Caesar; in contrast, his own work is in the best tradition of Christian propaganda.

style is evident. The figures are much larger and more powerful than any of Giorgione's. In one fresco, St. Anthony—who preached and died in Padua—grants speech to an infant, allowing the child to defend its mother against its father's charges of infidelity; in another, the saint restores life to a woman killed by her jealous husband; in the third, he heals a young man who has cut off his own foot to punish himself for kicking his mother. The characters in all these episodes are full of vitality. The virtuous mother listens to her child with gentle dignity; the terrified wife tries to fend off her husband's dagger; and amazement fills the faces of the spectators watching St. Anthony replace the severed foot.

Also in Padua, Titian departed from painting to design a woodcut that remains one of his most notable achievements, *The Triumph of Faith.* The wood blocks for this work were not actually cut until he returned to Venice, but its original design may have been based on a mural he is known to have painted on the walls of his room in Padua; the two at any rate deal with the same subject. Woodcuts, a popular art-form since the end of the 14th Century, had taken on new importance with the development of the printing press, and Venice had become a leading center of the young publishing industry. A German printer, Johann Speyer, had brought his presses to Venice in 1466, and cultured Venetians proved so eager for his books and prints that other publishing firms soon followed. The most famous was the Aldine Press, founded in 1490 by an Italian scholar, Aldus Manutius.

Manutius set himself the task of collecting, editing and publishing all the literature of ancient Greece, and he worked at it untiringly. "Those who cultivate letters," he wrote, "must be supplied with the books necessary for this purpose; and until this supply is secured, I shall not rest." He had chosen Venice as his base of operations because it was rich in money —and in Greek scholars. These men, fed and housed in Manutius's own home, served as his editors and translators. Meanwhile, Manutius himself cast his own type, mixed his own inks and commissioned the design of a semiscript typeface used ever since and known as italics. So pervasive was his influence that the members of Venice's newly formed humanist Academy spoke Greek, took Greek names for themselves and set a distinctly classical tone to the city's intellectual life.

By Titian's young manhood Venice had more than 200 presses and had published nearly 3,000 books on such subjects as philosophy, poetry, history and mathematics—more than four times the number of books published by its closest rival, Milan. Many of these volumes were handsomely illustrated with woodcuts by ranking artists; from their hands also came designs for woodcuts and engravings that could be printed in quantity, sold at relatively modest prices and hung on the walls of private homes. Among these artists, few could match the German Albrecht Dürer. Visiting Venice in 1505, Dürer made some trenchant observations on the Venetian art world. "There are plenty of fine fellows among the Italians who are becoming my friends," he wrote home, "people of sense and knowledge, excellent lute players, skilled pipers. . . . On the other hand there are also the falsest, most lying, thievish knaves in the whole world, I do believe, appearing on the surface to be good fellows. . . . My Italian friends warn me that I should not even eat or drink with their painters. Many of them are my enemies and they copy my work in the churches and wherever they can find it, and then they revile it and say that the style is not *antique* and so not good."

Titian's *Triumph of Faith* is more like a fresco than a woodcut. Indeed it may have been intended as a wall decoration for the homes of common folk who could not afford costlier art. Its size alone is remarkable—it was printed in 10 sheets some 9 feet long and 15 inches high. The subject is a procession much like those the Venetians so dearly loved in real life. Leading the procession are Adam and Eve, followed by characters from the Old Testament, followed in turn by Christ in a chariot pulled by winged symbols of the four Evangelists—a man, an ox, a lion and an eagle. Behind the chariot walk a host of Church notables, saints, martyrs and apostles. Titian was subsequently to produce other woodcuts, but none so monumental as *The Triumph of Faith*. In its unexpectedly bold, crude style, its strength and vigor, it not only established a precedent but also widened his reputation.

He returned to Venice in 1512, stopping off in Vicenza long enough to paint a fresco, *The Judgment of Solomon,* for the town hall—a work later inexplicably destroyed in the course of a remodeling of the building. Settled once more in Venice, Titian easily picked up where he had left off. Affluent citizens were still eager to acquire paintings by Giorgione, and they looked upon Titian as Giorgione's worthy successor. But the task of completing Giorgione's unfinished works only partially occupied his time; he also attracted commissions in which he was able to follow the stylistic inclinations that would earn him his own lasting niche. The style proved very different from Giorgione's in technique and in spirit. Titian laid his colors on more thickly; his brushstrokes were smooth and creamy; his figures, unlike Giorgione's idealizations, were frankly voluptuous.

It was in this period that Titian began to paint the lush portraits of women that are one of his special claims to fame. In one painting the subject clasps a bunch of flowers, in another a mirror; a third shows a curiously artless Salome holding the head of John the Baptist as if it were a new hat. Titian may have found his models among the celebrated courtesans of Venice. Some of these portraits are obviously pictures of the same woman—and all of them may have had a dual purpose. In addition to bringing pleasure to their purchasers, they may have been intended to publicize—and commend—the sitter's charms. Worldly Venice would have seen nothing amiss in the mixing of business and art. Many of its courtesans moved openly and easily in the highest levels of society. Their luxurious apartments were gathering places for noblemen and men of letters, and they entertained at banquets and concerts and balls as grand as any on the Grand Canal. Lavishly dressed, gifted in all the social graces, as adept in the art of conversation as in the art of making love, they adorned the city in such numbers and variety that in 1547 Venice published a directory of their names, prices and services.

Titian's paintings of these earthly love goddesses, so unlike the celestial Venuses of his contemporaries, culminated in a splendid canvas done to the order of the Grand Chancellor Niccolò Aurelio, the Venetian secretary of state. The painting has had various names—*Sacred and Profane Love, Artless and Sated Love, Beauty Adorned and Unadorned, Medea Harkening to Venus, Two Maidens at a Fountain (pages 70-71)*. Once its subject was thought to be drawn from classical literature, but apparently *Sacred and Profane Love* illustrates an incident in a popular "novel" of the day, the *Hypnerotomachia Poliphili* (The Strife of Love in a Dream with the Lover of Polia), by Francesco Colonna. The plan of the painting—the arrangement of the figures in the landscape—owes a lot to Giorgione, but in execution Titian's painting is not like Giorgione's at all. There is much more emphasis on texture, on the richness of the silks and satins; the colors are much more vibrant. Above all, the flesh of the nude is the flesh of a real woman, not an etherealized ideal. Venice may have wondered about Giorgione's meanings but it can have had no doubt whatever about Titian's.

By 1513 Titian's prestige had begun to match his aspirations, and he had to make a choice. He could stay in Venice and try for the position of first painter in his own city, or he could seek renown in Rome, which under the Medici Pope Leo X was Italy's liveliest center of patronage of the arts. Each alternative had its advocates. One of Titian's friends, the poet Pietro Bembo, had recently gone to Rome to be secretary to the new Pope, and he urged Titian to follow him. Bembo had been something of a personage in Venice—an art connoisseur, a classicist, author of some lofty if stilted verse published by the Aldine Press, and a personal friend of Isabella d'Este. Mindful of the credit that would accrue to him if he could succeed in luring Titian to Rome, Bembo must have used all his powers of persuasion, and Titian must have been tempted.

But there were equally strong voices urging Titian to stay home. His influential patrons in the Pesaro and Barbarigo families did not wish to lose his services. In addition, there were appeals from such men as Andrea Navagero, a staunch classicist and a member of the Venetian Academy. Their arguments prevailed. On May 31, 1513, Titian presented a petition to the Venetian Council of Ten, asking to be appointed supervisor of the decorations of the Great Council Hall and proposing to paint for the Hall a battle scene—of whatever battle the Council might designate. Beneath its mock-modest language the petition was in effect a bid to be named official state painter of Venice:

> I, Titian of Cadore, having studied painting from my childhood, and wishing for fame rather than profit, desire to serve the Doge and Signori rather than His Highness the Pope and other lords who in the past, and now once again, have asked me to enter their employment. I therefore propose, if it should be possible, to paint in the Hall of Council, beginning, if it should please your Highnesses, with the canvas of the battle on the side towards the Piazza, which is so difficult that no one has yet had the courage to try it. I am willing to accept any payment thought proper, but thinking only of honor and asking no more than a moderate amount, I beg leave to have the first broker's license for life that shall be vacant in the Fondaco dei Tedeschi, irrespective of all promised reversions of said license, and on the same conditions or with the same exemptions as are conceded to Messer Gian Bellini, that is, two youths as assistants to be paid by the Salt Office, and all colors and necessary materials. In return, I promise to do the work above named with such speed and excellence as will satisfy your Lordships, to whom I beg to be humbly recommended.

In asking to be granted the same favors as Giovanni Bellini, Titian was claiming to be the equal of his old master and was proposing that he be named Bellini's successor. For 30 years Bellini had reigned unchallenged as the leading painter of Venice. Alone or with the help of his brother, he had produced some eight or nine murals for the Great Council Hall. He had also supervised the work of other painters working in the Hall, even the work of so eminent an artist as Carpaccio. Now, however, Bellini was in his eighties; his powers were declining and his potential successors were few. None of Bellini's assistants showed great promise, and among the more established painters most were either aging or too busy with private commissions.

Titian's self-confidence and his growing reputation must have impressed the Council of Ten. Within a week the councillors accepted his petition "with all the conditions attached to it" and ordered the Salt Office to furnish him with money for materials and assistants. Titian moved into a new studio in the quarter of San Samuele, a property once owned by a Milanese nobleman and later taken over by the Venetian republic to provide housing for artists and architects working on state projects. In these new quarters and with two assistants, Titian immediately began to make sketches for his proposed battle scene for the Council Hall. But things did not go altogether smoothly; Titian's action had aroused the jealousies and antagonisms that Albrecht Dürer had earlier encountered in the Venetian art community. Within a year, the Council of Ten suddenly reversed itself and ruled that Titian could not have the first vacant broker's license in the Fondaco dei Tedeschi after all; he would have to wait his turn, and in the meantime the state would not pay the salaries of his assistants. Giovanni Bellini is generally credited with instigating this move, egged on possibly by other painters who felt that they had prior rights to the Fondaco license.

Titian waited eight months to reply to this new ruling—perhaps because he had enough private commissions to keep him busy. Then, in November 1514, he went before the Council and complained of being the victim of a conspiracy. His battle scene, he said, would long since have been finished if the Council had not cut off his allowance. He protested that if he did not get the broker's license in the Fondaco, he would surely starve. Would the Council, he asked, give him, if not the first vacant license, at least the license held by Bellini when the venerable painter died? The Council assented. Moreover, wishing to see some progress on the proposed battle scene, it ordered the treasury once again to furnish Titian with money for paints and assistants. It even agreed to mend the leaking roof in Titian's studio—providing the bill did not exceed six ducats.

The squabbling, however, did not end. In December 1515 one Council member charged angrily that the cost of decorating the Council Hall was becoming excessive. Enough had already been spent on it, he said, to adorn all of the Doge's Palace. He demanded an accounting of all expenses to date, and an investigation into the status of all the paintings—and painters—involved in the Hall project. The next day another councillor pointed out that at least two of these paintings, although still only in sketch form, had already cost the government 700 ducats—and that he knew of other painters who would have provided the same work for much less. The upshot of this storm was a resolution discharging all the artists at work on the Hall and ordering the Salt Office to pick a single man to finish the project and bargain with him for each picture.

Somewhere in the background of this explosion may have lurked Titian's wily hand; the reference to "other painters who would work for less" may have come from his partisans on the Council. In any case Titian seized the opportunity to approach the Council with a brand new offer. He would, he said, paint his battle scene for the sum of 400 ducats, plus the salary for one assistant, plus 10 ducats worth of colors and the promise of Bellini's license in the Fondaco when it became available. The Council knocked 100 ducats off the price and accepted the offer. Within a year, on November 29, 1516, the diary of Marino Sanudo, a Venetian gentleman who chronicled the daily events of Venice, noted that "this morning we were told of the death of Giovanni Bellini, that fine painter . . . famous throughout the world. . . ." Soon afterwards Titian fell heir to Bellini's license, becoming to all intents and purposes the first painter of Venice—an accomplishment unprecedented for an artist not yet 30.

Titian's Master

Around 1507 the Venetian master, Giorgione of Castelfranco, took Titian of Cadore into his studio as an assistant. Out of their relationship came several of the most brilliant collaborations in the history of painting and some of the most perplexing problems in art scholarship.

The few years of Giorgione and Titian's active partnership proved a crucial and rewarding time for both artists. It was then that Giorgione's highly personal style attained maturity. His most characteristic works were private, dreamlike visions of man immersed in the natural world. And, although his own thoughts on the matter are unknown, it seems likely from looking at these pictures that Giorgione came to regard painting primarily as a medium of self-expression and only secondarily as a vehicle for storytelling. Certainly, his contemporaries often struggled to discover the precise meaning of some of his pictures, but they unfailingly admired his extraordinary magical "landscapes."

Titian worked with the master on several canvases at the same time he was developing his own distinctive style. After Giorgione's untimely death in 1510—he was in his early thirties—Titian altered and completed some unsold works. This close collaboration, and the almost total lack of documentary evidence, has made it extremely difficult to determine where the hand of Giorgione left off and the hand of Titian took over. If the answer is to be found, it will be in a close study of the painters' unique styles.

65

The essence of Giorgione's subjective art is contained in these two pictures, both of which reveal his preoccupation with beauty and with communicating the mood of a particular moment rather than narrating a tale. In *The Tempest* the nursing mother and the young man exchange not even a glance that might reveal some personal relationship between them. Is he her lover, a passing soldier, a protecting saint? Situated in an enigmatic setting, they are figures of fantasy, frozen in separate reveries. The pensive figures in *The Three Philosophers* (a widely disputed title) appear to be specific individuals, yet they cannot be identified. One scholarly theory suggests that they represent the Magi who paid homage to the Infant Jesus. But, for Giorgione's purpose, the identities of

Giorgione: *The Tempest*, c.1505-1508

all these figures seem irrelevant. If there is a theme in these pictures, it is the drama of all nature, in which each element—man, tree, rock, sky—plays an important part. A close look at Giorgione's unconventional technique reveals how he accomplished this unity. Working without the usual meticulous preliminary studies, he sketched his composition loosely and directly on canvas. His foreground figures, modeled with soft contours and hazy outlines, seem distant. By contrast, he painted background objects—the tree, hills and buildings in each scene—with a clarity of form and color that makes them seem near. Finally, Giorgione achieved the force of his drama by suffusing each scene with pure color, at times bright and electrifying, often sombre and mysterious.

Giorgione: *The Three Philosophers,* c.1506–1508

Titian was less interested than Giorgione in nature per se. His tastes were for human drama rather than poetic tableau, for worldly realities rather than spiritualized beauty. Even early in his career Titian began to show this preference for emotional and physical action in works like the drawing of domestic passion shown on page 52. In contrast with Giorgione's intimate, allusive art, Titian's paintings tended to be grand, forceful and substantial.

Sacred and Profane Love was painted after Giorgione's death, when Titian was on his own. The allegory is clear and dominant, enacted before a classical marble frieze of Venus and Adonis. Seated at the left is a richly dressed female figure with jewels at her waist and flowers in her lap, who represents the love of worldly things; at her right

Sacred and Profane Love, c.1515

amor celestis, untrammeled by earthly possessions, holds the burning lamp of Divine Love. Where allegory is present in Giorgione's pictures, it is ambiguous; in Titian's the symbolism is explicit. Even his technique is bold. Titian uses a firmer brushstroke and models his figures with a clearer outline, creating an impression of robust vitality. The foreground here is filled with eye-catching details:

opulent draperies, ripe cherries, lustrous metal, hard stone, soft flesh. By emphasizing the precise physical quality of such things, Titian has subordinated the landscape background in a way that Giorgione never would. He has not eliminated nature—there is much pastoral detail in the scene—but it is clearly subservient to the radiant dual image of feminine beauty that Titian presents.

Music fills this pastoral scene; the mood of quiet listening, the languid rhythm of a hot summer afternoon in the open air is implicit. Giorgione is known to have had a passion for music, and the lyrical atmosphere of this painting is consistent with his other works. The tranquil beauty of the nudes, the softened contours of their well-fleshed bodies, also mark them as creatures of Giorgione's romantic imagination. Their actions seem dreamlike and disengaged; the girl at the well pours water absentmindedly, the negligent flutist gazes not at her companions but into the distance.

There is a disturbing note here, however —a note of inconsistency. The young men are more solidly delineated by the painter's brush. They seem more animated and more individualized than are Giorgione's usual subjects. Even the landscape is unlike his active, integrated natural settings.

Here is the heart of the matter, the center of a controversy that has plagued art historians: who painted this picture? Even the title, given here in French because the picture is owned by the Louvre, is disputed. Known widely as *Fête Champêtre,* or "Country Feast," the picture has had many other names, including "Pastoral Symphony" and "Fountain of Love."

The answer to the question may never be found, but in the analysis of style related above lies a most persuasive explanation: Giorgione selected the theme and perhaps painted the women and certain other details; Titian painted the men, much of the landscape and perhaps was responsible for integrating the entire work. It would seem that both men worked on the painting, perhaps even at the same time.

Giorgione and Titian: *Fête Champêtre*, c.1505-1510

Another great masterpiece, *The Concert*, remains in the limbo of disputed authorship. It has been suggested that Giorgione and Titian both worked on the picture—not simultaneously, as in the *Fête Champêtre*, but separately and perhaps years apart. The speculation runs this way: in the 16th Century, painting was a laboriously slow and meticulous process (Titian is known to have worked on canvases for years), and at Giorgione's death a number of unsold canvases were left in his studio. Titian, as his foremost associate, may have had access to those paintings, and either out of homage to his late master or simply as a practical matter, he may have altered or finished certain of the works. The effort of two hands on this picture seems clear. It is somewhat easier than in *Fête Champêtre* to detect the difference in styles because only three figures are represented and the indoor setting leaves fewer particulars to be examined.

What first strikes the eye is the apparent discrepancy between the way the monk at the harpsichord and the other figures are painted. The precise outline of the monk's left cheek, the drama of his flashing, sidelong glance, the dynamic attitude of his fingers poised over the keyboard express an energy and vitality that suggest the hand of Titian. By contrast, the shadowed listeners, whose very characters are ambiguous—is the feather-hatted figure at the left a man or a woman?—evoke the softly lighted, poetic style of Giorgione. Here again the theme is music, but now the staccato performance of Titian can be compared with the lyrical tone of Giorgione.

Undoubtedly there were other uncompleted works among Giorgione's effects that may subsequently have been completed by Titian or by members of the master's studio. But in this picture, and the others that he touched with his spirit, the genius of Giorgione shines through.

Giorgione and Titian: *The Concert*, c.1515

IV

The Painter
as Poet

The broker's license that the Venetian republic awarded Titian in 1517 carried with it certain obligations. In return for the comforts of a regular income and a tax exemption of some 20 ducats a year, he was to provide the state from time to time with paintings of various dignitaries and events. As each new Doge came to power, Titian was supposed to do a life-sized portrait of him and also a devotional picture, showing the Doge paying homage to the Virgin. In addition, the government expected Titian to produce his promised battle scene for the Great Council Hall. But he took his time about meeting these responsibilities. He was deep in work—but not work for the state. Using one excuse after another, he managed to put off the battle painting so successfully that 20 years later the government was still demanding furiously that he finish it.

Titian's neglect of his official duties has been explained in several ways. Some of his defenders point out that the wall space assigned for the battle scene could hardly have been inspiring: since the wall faced south, the sunlight flooding through its windows must have thrown the space between them into semi-darkness. Another argument in Titian's behalf is that he may have slighted his own tasks because he was busy completing a painting left unfinished by Giovanni Bellini, commemorating the reconciliation arranged by Venice in 1177 between Pope Alexander III and the Holy Roman Emperor Frederick Barbarossa. The likeliest explanation of all, however, may be that Titian simply found his private commissions more interesting and more lucrative.

One of these commissions was for an altarpiece *(left)* that filled its patrons with misgivings while it was in progress but proved to be a revolutionary work for 16th Century Venice. Around the time Titian received his broker's license, he was asked by Father Germano, prior of the Franciscan monastery of Santa Maria dei Frari, to produce a painting to place behind the high altar of the Franciscans' church. The frame for this altarpiece was already on order, a tremendous carved marble arch more than 22 feet high and over 11 feet wide. Within it, the Franciscans wanted a painting of an Assumption of the Virgin. What they had in mind was an Assumption rather like the one Giovanni Bellini had painted for the Church of St. Peter Martyr on the island of Murano—a gentle, thoughtful Madonna on a soft cloud, suspended above an assembly of identifiable saints backed up by a landscape full of interesting pictorial detail. What they got was an Assumption with no background landscape at all, a painting in which heroic figures enacted a breathtaking drama against a vast emptiness of space.

Titian worked on the Frari altarpiece off and on for two years. From time to time Father Germano came to peer nervously over his shoulder and comment on the excessive size of the foreground figures. They were indeed disproportionately large compared to the figure of the Virgin, but Titian had no easy way of explaining his purpose, so he put off his questioner as one would put off a child, simply saying that a big church needed big figures. In fact, the disparate size of the figures was meant to intensify the impact of the work. Titian was not painting figures in a realistic landscape, seen as if glimpsed through a window; he was re-creating an emotional experience. He intended the eye to be engaged by the tense, passionate figures of the Apostles in the foreground, to ascend from them to a vibrant figure of the Virgin, and to rise from the Virgin to a benign God, aloft in the distant empyrean just beneath the top of the arch of the marble frame.

Titian's *Assumption* was not the comfortable retelling of a familiar tale; it was the dramatization of a wondrous event. It carried the viewer into the very heart of a religious experience and made him a participant. The Franciscans were not at all sure that this was how a religious theme ought to be treated. Rumors flew that they were appalled, that Titian had threatened to keep the painting for himself, that he had forced Father Germano to apologize for his doubts. But when the painting was unveiled in March 1518, crowds of Venetians drawn by the advance publicity came to see it—and were enchanted. The envoy of the Holy Roman Emperor offered to buy it on the spot. By the end of the day the timorous Franciscans must have been reassured, for there was no more talk of unsuitability. In fact, the *Assumption* graces the Frari church today.

Titian's religious paintings matched the spirit of his age and his city. They were as worldly, as sensuous and as lushly colored as Venice itself—and Venice, with its gilded and painted palaces and its liquid sunsets, was the most colorful city in Italy. The Frari Virgin is not some ethereal creature but a strikingly vigorous woman. Her garments are in sparkling tones of red and blue; her draperies have a lively swirl. Painting her was not an act of piety but an act of art by a man sensitive to the classical currents of his time: the Virgin, majestic and larger than life, instantly conjures up the sculpture of ancient Greece and Rome. Venetians had never seen her like. New commissions poured in on Titian, and he began to experience the flattery of imitation.

Titian accepted both the homage and the new work with enormous gusto. The year after completing the Frari altarpiece he began another painting for the same church, almost as large and every bit as daring. The powerful Pesaro family commissioned a votive picture to be hung above a side altar; it was to show members of the family kneeling before a Madonna and Child. The customary format in paintings on this theme called

In the same magnificent Venetian church—Santa Maria dei Frari—that houses two of Titian's most glorious religious paintings (the *Assumption* shown on page 76 and *The Pesaro Madonna* illustrated on page 114) is the artist's tomb *(above)*. An ornate and rather clumsy memorial erected over it in 1852 pictures the bearded master seated between Nature and Knowledge and flanked by figures representing Painting, Engraving, Sculpture and Architecture. Crowning the triumphal arch above him is the Lion of St. Mark (symbol of Venice) with the Habsburg shield. Titian is honored in the church along with a variety of other artists and noteworthy Venetians of the past, including two important Doges.

for the Madonna to be placed in the center and flanked by a balanced arrangement of worshipers—a composition that suggested serenity and order. However, Titian's painting *(page 114)* set the Madonna off to one side, with the other figures clustered about unevenly, resulting in a much more complex sort of balance than that of the traditional even-sided pyramid. The viewer's eye is compelled to travel back and forth in a zigzag line between various points of emphasis. The effect is not one of calm, but of almost agitated movement.

The Pesaro Madonna is also remarkable for the realism of its portraiture. The kneeling worshipers are properly grave, but the gravity is that of worldly men to whom pious pursuits are more a matter of duty than of conscience. The attention of one young Pesaro has wandered from the solemn occasion: a boy, his face framed by the dark cloaks of his elders, looks directly out at the viewer. Among the more illustrious members of the family in the painting are Benedetto Pesaro, once admiral of the Venetian fleet, and Jacopo Pesaro, Bishop of Paphos, the same Jacopo whom Titian had painted some 15 years before, depicting his presentation to St. Peter by the Borgia Pope Alexander VI. The toll of the intervening years shows on the bishop; his face in *The Pesaro Madonna* is lined and leathery. But evidently he still found it prudent to acknowledge his debt for past favors by the Borgias, for the Borgia coat-of-arms appears in the devices of the standard fluttering above his head.

While still at work on *The Pesaro Madonna,* Titian undertook another altarpiece for the Church of St. Francis in Ancona, a seacoast town about 200 miles south of Venice, and yet another altarpiece for the Church of St. Nazarius and St. Celsus in the city of Brescia, commissioned by the papal legate to Venice. This painting, a *Resurrection,* is divided into five panels, in one of which the kneeling patron himself is seen, plump, hooknosed and slightly cynical. But the most striking panel in the altarpiece is one that portrays St. Sebastian, pierced by an arrow and tense with pain. Titian thought it the best thing he had ever done and showed it to his friends in a private viewing in 1520 before the entire altarpiece was completed. Word of its beauty reached the ears of the Duke of Ferrara, who tried to acquire it for his personal art collection. The Duke withdrew his offer when he realized that it might offend the papal legate, but not before Titian, in a bargaining session with the Duke's agent, proudly insisted that while the price set for the altarpiece as a whole was 200 ducats, the figure of St. Sebastian alone was worth that sum.

A commission for a fourth altarpiece during this period brought Titian in contact with one of the strangest artists of the Renaissance. The altarpiece was an *Annunciation* for one of the chapels of the Cathedral of Treviso, which was simultaneously being decorated with frescoes by a painter best known to history by the nickname Pordenone, after the town he came from. Pordenone was a bit older than Titian and much less talented. He painted in a powerful but crude style that caused Vasari to suggest that he was self-taught. More likely, the style was a matter of personal choice rather than inadequate training, for Pordenone's work continued to exhibit a kind of peasant burliness to the end of his life. One of his finest paintings, a *Crucifixion* for Cremona Cathedral, is ferocious in mood and full of tortured shapes, much closer in spirit to Goya, several centuries later, than to the classical temper of his own time.

Among Pordenone's outstanding characteristics was his speed. He was said to have started and finished his first work, a *Madonna,* while the man who ordered it was attending Mass. So swiftly did he complete a fresco for a house in Mantua that people believed he was in league with the devil. Indeed there was a demoniacal quality about Pordenone. He wandered impatiently from town to town, never staying anywhere long; he married three times, returning each time to his home town to choose his bride; he quarreled and brawled and was quick to take offense. When his father died, Pordenone wrangled with his brother over their father's estate, at one point reportedly engaging his brother in a duel, at another point picking a street fight that caused the death of a friend.

Predictably, Pordenone's relations with Titian were erratic. When the two worked together at Treviso they were apparently on the best of terms. Titian defended Pordenone against an angry patron who claimed that Pordenone had overcharged him, assuring the man that on the contrary he had gotten a bargain. Pordenone once exclaimed, on seeing a Titian altarpiece, "I believe that this is real flesh, not color!" But in 1528 an event occurred that soured the friendship. In Venice, the Scuola of St. Peter Martyr invited artists to compete for a commission for an altarpiece to its patron saint. Among the entrants were Titian and Pordenone, and Titian won. Thereafter Pordenone hated Titian and went around claiming that Titian had threatened his life—a most unlikely action for the man Vasari described as being "of unvarying courtesy and affability."

The *St. Peter Martyr* altarpiece has often been called Titian's finest religious painting. Artists of his own time praised it extravagantly, and many later artists copied it. Vasari hailed it as "the best and most perfectly finished, as well as the most renowned of any that Titian ever executed," and Titian's friend, the writer Pietro Aretino, claimed that at the sight of it the goldsmith Benvenuto Cellini and the sculptor Tribolo were "converted into statues of astonishment." The painting showed St. Peter being attacked by an armed man while a companion flees in terror; at the top, two angels descend, bearing the palm of martyrdom. Years later, the connoisseur Francesco Algarotti marveled at the trees and plants in the painting, so exquisitely rendered that "a botanist would have much ado to keep his hands from them," while Sir Joshua Reynolds was inspired to observe that Titian's handling of "color and light and shade" in the *St. Peter Martyr* made him "both the first and greatest master of this art." Unfortunately, the altarpiece fell victim of its own fame. In the 18th Century it was removed from its original site in the Church of St. John and St. Paul and was sent to Paris; sometime after that a wealthy agent tried to buy it. Alarmed, the Venetian government ordered that the painting never again be taken from the Church of St. John and St. Paul. In 1867 a fire swept the church, and the painting went up in the flames.

Titian's altarpieces introduced a new element of drama into devotional art. To the delight of his humanist patrons, he employed this innovation in his secular paintings as well. In his hands the *poesie,* the form of idyllic painting so popular with Renaissance princes, became a Dionysian revel in which gods and goddesses and tumbling cherubs invited men to share the joys as well as the classical refinements of the antique spirit. In the same eventful year that Titian fell heir to his broker's license and acquired the commission for the Frari altarpiece he began painting, for a single patron, a series of *poesie* on a single theme; they are among his most enchanting paintings.

The patron was the Duke of Ferrara, Alfonso d'Este, brother of the celebrated Marchioness of Mantua, Isabella d'Este, and husband of the notorious Lucrezia Borgia. Alfonso had inherited his father's title in 1505, and also his considerable fortune—and he knew precisely how he wanted to spend it. He wanted to free his little duchy from Venetian rule, and he wanted to rival his sister as a collector of art. His failure at the first was matched by his success at the second. In 1508 he joined the League of Cambrai in its war against Venice and promptly lost as much territory to the greed of his allies as he did in military actions. In 1511 he refused to join a second League against the French formed by the Pope and the Venetians; caught between Venice and Rome, he found himself in the position of a tasty bone between two hungry dogs. By 1513, when the warring parties finally settled down to an uneasy peace, Alfonso had lost so much money and property that he had had to pawn the family jewels for ready cash.

None of these misfortunes deterred Alfonso from the pursuit of his other goal, to win renown as a connoisseur and collector of art. Soon after his father's death, he had begun to enlarge and redecorate his castle at Ferrara, a medieval fortress built sometime after 1385. Among the improvements was a series of small rooms built over a passageway between two parts of the castle. The rooms were sheathed in marble of such dazzling whiteness that they came to be called the alabaster chambers *(page 88)*, and they were the Duke's pride and joy. Around 1513 he decided to decorate these chambers with paintings in the classical style, in imitation of similar paintings that hung in the private study of his sister Isabella in Mantua.

Among the sights that surely thrilled the young Titian on his arrival in Venice was the lovely little church of Santa Maria dei Miracoli. Completed only a few years earlier (about 1489), the church was built to house a miraculous image of the Virgin. Designed and decorated by Pietro Lombardo and his sons and sheathed in richly colored marble, the building itself is a jewel of Venetian architecture, expressing a Byzantine love of surface splendor and an early Renaissance interest in classical forms. In the relief below, one of innumerable carvings within the church, a pagan mermaid is intertwined with angelic cherubs. Titian himself later adopted such imagery in his paintings.

Isabella's classical paintings belonged to a genre much favored by cultured Renaissance noblemen who in painting, as in the other arts, idolized the accomplishments of ancient Greece and Rome. The names of long-dead painters like Protogenes and Apelles fell as easily from their lips as the names of Raphael and Michelangelo. Even though no examples of the work of the ancient artists survived, admiring tales of their technical skill were circulated. One such story told of a competition between two Greek painters, Zeuxis and Parrhasios, to determine who was the better at his art. Zeuxis, it was said, painted grapes so real that birds flew down to peck at them, but Parrhasios painted a picture of a curtain that Zeuxis mistook for a covering over a painting and tried to draw aside. When he discovered his error Zeuxis acknowledged defeat, conceding that while his own picture had fooled the birds, Parrhasios's had fooled a man.

To own a painting done by one of these artists of antiquity was of course impossible, so Renaissance art collectors settled for what they fondly believed to be the next best thing. They commissioned artists of their own day to paint pictures for them in the classical style. Some of these classical paintings were actually based on descriptions of the work of such men as Apelles and Parrhasios. More often they were simply illustrations of incidents in stories based on Greek and Roman mythology—as told by Greek and Roman writers, or by Renaissance writers with a talent for imitating the Greek and Latin originals. Whatever their sources, Renaissance painters took such pains to duplicate on canvas precisely what they read in books that it is often possible to pinpoint the exact passages on which their paintings were based.

This was the sort of painting that Alfonso d'Este had in mind for his alabaster chambers, and the theme he chose was the delights of love and wine. The first picture he commissioned was a *Bacchanal* by Dosso Dossi, illustrating a passage in Ovid's *Fasti,* a lengthy retelling in verse of the mythical stories behind the feast days of the Romans. Then Alfonso acquired Giovanni Bellini's *Feast of the Gods,* also inspired by the *Fasti*—a painting originally intended for Isabella d'Este's collection until Alfonso snared it away. With room for perhaps three more pictures to accompany these two, Alfonso dipped into the classics in search of suitable subjects. He found two he liked in the *Imagines* by the Greek scholar Philostratus the Elder (a book he borrowed from his sister's library), and a third in Ovid's *Fasti.* Philostratus' book, written in the Third Century, describes 64 antique paintings the author claimed to have seen in an art collection in Naples. The two works Alfonso chose to have "reproduced" were a painting of a shrine to Aphrodite in a garden full of cupids and a painting of a celebration honoring Bacchus on the island of Andros. The passage from Ovid that provided Alfonso with his third subject tells of Bacchus's meeting with the Cretan princess Ariadne on the island of Naxos.

Alfonso's first choice to do the Aphrodite was the painter Fra Bartolommeo, a Dominican monk with whom the Duke had been negotiating for several religious pictures for his wife. Death, however, saved Fra Bartolommeo from having to execute a subject so unsuited to his sacred calling. For the Bacchanal on Andros, Alfonso approached Raphael, then the darling of the Italian art world; the Duke had met Raphael on a visit to Rome and had been promised a painting by him. But Raphael refused to touch the Bacchanal. He had heard, he wrote, that the commission had also been offered to another painter, and he preferred therefore to do something else for the Duke, on another theme. For more than a year, Alfonso continued to urge the Bacchanal on Raphael, and Raphael continued to reject it. Finally, when it became obvious that his appeals were fruitless, Alfonso commissioned all three works from Titian.

As it happens, Titian was already in Alfonso's pay. For seven weeks in 1516 he had stayed at the castle in Ferrara with two assistants and had received, according to the castle records, an allowance of "salad, salt meat, oil, chestnuts, oranges, tallow candles, cheese, and five measures of wine each week." But there is no mention of the services he rendered in return. A year later Titian wrote the Duke from Venice to say that he had gone "without delay to the well of which your Excellency had written and made a sketch of it," and that, not wishing to send this sketch alone, he was including "another with it, of a well after the fashion of this country." Then, with the kind of obsequious flourish that business-minded Venetians found so useful, Titian tacked on this sentence: "I am entirely at your command should the drawings be considered unsatisfactory, and am ready to furnish others; because having given myself body and soul to your Excellency there is no pleasure I esteem so great as to be worthy of serving when and where your Excellency may think me fitted to do so." Of the "wells" he had sketched, there is no further information.

In April 1518 Titian wrote Alfonso acknowledging the receipt of canvas and frame, along with a proposed subject for a painting. This Titian agreed to finish "at the appointed time." Three weeks later Alfonso's agent in Venice, Jacopo Tebaldi, wrote him confirming the arrival of "a sketch of a figure and complementary notes," which Alfonso had sent for Titian's guidance, and asking the Duke to tell him exactly where the painting was to be hung. Both these letters apparently referred to the so-called *Worship of Venus (page 91)*, Titian's painting of the shrine of Aphrodite, which, true to the description in Philostratus's book, is a positive mass of cupids frolicking around a statue of Venus attended by two nymphs. "Here run straight rows of trees with space left free between them to walk in," writes Philostratus, "and tender grass borders the paths, fit to be a couch for one to lie upon. On the ends of the branches apples golden and red and yellow invite the whole swarm of Cupids to harvest them." And so it is in Titian's painting.

A year or so later he produced the *Bacchanal of the Andrians (page 90)*. But in the interval Alfonso found Titian something of a trial; despite professions of his desire to be of service, he was dragging his feet. "We thought that Titian, the painter, would some day finish our picture but he seems to take no account of us whatsoever," Alfonso wrote Tebaldi in September 1519 when the *Bacchanal* had not yet arrived. "We therefore instruct you to tell him instantly . . . that he must finish it under all circumstances or incur our great displeasure." Titian then wrote soothingly that he would shortly bring the painting to Ferrara and finish it there. Toward the end of October he kept his word, traveling to the ducal estate by barge up the river Po, accompanied by his canvas. The Duke must have thought the wait worthwhile: on viewing the completed *Bacchanal,* the most glorious celebration of drunkenness ever seen, he promptly awarded Titian several new commissions.

At this point Titian had far more work than he could possibly handle, even for a man of his immense energies, and importunate patrons pressed him from every side. Yet he went his own serene way, filling orders in his own good time with a growing independence of spirit. In 1518 indignant Venetian officials had reminded him that he still had not completed the long-promised battle scene for the Great Council Hall and had threatened to turn the project over to someone else—to be finished at Titian's expense. Titian, busily at work on Alfonso's *Worship of Venus,* had paid them no heed and had compounded the injury by accepting orders for several new altarpieces that same year. In 1521, leaving behind a number of unfinished canvases, he went off to the town of Conegliano to paint a fresco for the façade of the Scuola of Santa Maria Nuova. Later that year he returned to Venice to start a portrait of the new Doge, Antonio Grimani, and at the same time to work on the promised altarpiece for the Church of St. Nazarius and St. Celsus at Brescia, as well as on the *Bacchus and Ariadne* for Alfonso's alabaster chambers.

In December 1521 the Duke wrote Titian inviting him to spend Christmas in Ferrara and to bring the painting with him. Titian declined politely but firmly, but the Duke's invitations persisted. In January 1522 Titian pleaded illness; in June he claimed to be unsatisfied with some parts of the *Bacchus and Ariadne.* In August Tebaldi, Alfonso's agent, called round at Titian's studio to examine the work and was astounded to find that it contained only two figures and a leopard-drawn chariot. The rest of the canvas was blank. Titian assured Tebaldi that it would be finished in two weeks once he got down to it, but in October he was still putting the Duke off, protesting that he could not bring the canvas to Ferrara to finish it because there were no good models there.

Tebaldi, in a letter to the Duke, assured him that Titian was working as hard as he could but that he was having to divide his time between the *Bacchus and Ariadne* and the battle painting for the Great Council Hall— for the government had again threatened him, this time with the loss of his broker's license. In January 1523 Titian finally set out for Ferrara with the Duke's completed painting—but even then he did not go there directly. On the way he stopped at Mantua to look at Isabella d'Este's art collection and to discuss some possible work with Isabella's son, the new Marquis of Mantua, who had inherited his father's lands and title four years earlier. It was not until February 1523, three years after Alfonso d'Este had commissioned the *Bacchus and Ariadne,* that he saw it in place in his alabaster chambers.

During this visit to Ferrara and in subsequent visits Titian apparently altered *The Feast of the Gods,* which Giovanni Bellini had executed about a decade before. In an attempt to make Bellini's painting more harmonious with his own paintings for the Duke, he repainted Bellini's background, added tokens of office to some of the gods and subtracted some of the nymphs' clothing. But these modifications did little to bridge the gap between two quite different kinds of painting. Titian, following in

Giorgione's footsteps, was seeking to provide his viewers with something more than a careful re-creation of a literary theme. He was attempting to recapture the classical spirit of the story or poem that was his source material. His *poesie* may look like simple fantasies or invitations to sensual delights, but in fact they are reproductions in paint of the specific emotional states that animate the words of the originals. What better way could a painter serve the poets of antiquity than by becoming himself a poet in paint?

That painter and poet had parallel functions was an idea that often occupied the attention of the Renaissance. Lodovico Dolce spoke of painters as mute poets and poets as articulate painters, and went on to observe that poets represented not only what the eye saw but what the mind comprehended. Similarly, painters, although they could not show sensory things like the coldness of snow or the sweetness of honey, could nevertheless indicate the sensations aroused by these things through facial expressions and poses. Vasari was alluding to a similar concern with emotional content when he said that the figures in a painting by Gentile Bellini were "in very fine attitudes." And many centuries later the English painter James Northcote ascribed a similar concern to Titian, in writing that one of Titian's women showed, "by her downcast look and her arms folded on her breast the agitation of her mind."

The artist's use of his feelings as well as his technical skills was sometimes a conscious act and sometimes not. "To paint a beauty," Raphael once wrote, "I make use of a certain divine form or idea which presents itself to my imagination." Conscious or not, the transition from literal painting to expressive painting was a step of immense daring. Between the tranquil Madonnas of Giovanni Bellini and the dramatic Madonnas of Titian there was a world of difference.

The means by which Titian achieved this difference was color. Never in the history of painting had it been so rich, so vibrant, so self-sustaining; never had it been put to canvas so freely and confidently. To Titian, moving into his prime as a master, color became the very essence of painting—the only way to establish a mood, communicate a feeling, awaken a response in the viewer. He would employ color in bold harmonies or subtle nuances or even deliberate discordances, but always it would serve to give pulsing life to his works.

Venetians, ever in love with the sensuous, reacted to Titian's color with predictable relish. Some non-Venetians were put off; Vasari, for one, noted rather sourly that a painter could "hide his lack of design beneath the attractions of coloring." But Vasari was trained in the artistic world of Florence, where line and form were primary considerations and color secondary; in any case, such doubts as he raised were increasingly in the minority. In the 18th Century the influential Sir Joshua Reynolds wrote of Titian: "By a few strokes he knew how to mark the general image and character of whatever object he attempted; by this alone conveying a truer representation than . . . any of his predecessors who finished every hair." It was given to Titian to be the first artist to make of color an entity unto itself; and in so doing he opened a path that generations of painters have followed down to the present day.

The Princely Patrons

It was a unique coincidence of history that Titian's evolving style—colorful, sensuous, immediate—perfectly matched the tastes of such wealthy and influential patrons as the ruling families of Mantua, Urbino and Ferrara. To be sure, like painters long before and after him, Titian painted what was ordered and not what his fancy dictated. But it seems that each picture he made, whether religious, mythological or contemporary, satisfied both his own ideas about what a painting should be and his patrons' notions of what they wanted. This was so because their attitudes toward life were strikingly similar. Both artist and patron were involved in a new humanist tradition that saw man as the measure of his world, and each felt that this world should be made to embody beauty and esthetic pleasure.

Titian's patrons wished to live in beauty and splendor, and often did so beyond the limits of their fortunes. In addition to paintings, they commissioned lavish jewelry, bought sumptuous fabrics and filled their palaces with finely crafted furnishings. Educated in the great events and works of the past, they found in the classical world a sympathetic model for existence. At the same time they were as little bound by the disciplines of antiquity as they were by the orthodoxy of religion. In all, they were men and women of many parts who desired to experience life through art and to make living itself an art. They were aided by the most brilliant painter of the time—Titian.

The desire of 16th Century Italian princes to live surrounded by art is epitomized by the Hall of Mirrors in the Gonzaga palace in Mantua. Like most such rooms of the period, the style of its architecture and the theme of its embellishments are classical. The impressive ceiling frescoes, designed by Giulio Romano, portray the chariots of the sun and moon and the thundering frolics of Olympian gods.

Antonio Lombardo: *The Forge of Vulcan (center)* and decorative reliefs from the alabaster chambers, c.1508

Alfonso d'Este *(opposite)* was a member of one of the most powerful Italian families, for generations masters of Ferrara, chief among the smaller states. The sumptuous court that flourished under Alfonso set the tone for much of Italian noble life. The man himself was an earthy yet princely patron with a mania for collecting. He turned every element in his environment into an object of delight and display. His most prized "jewels" were alabaster chambers— so called because of the brightness of their marble sculptures—designed as his private retreat. The reliefs at the left were made for the chambers by Antonio Lombardo; the decorative borders recall Imperial Rome;

Portrait of Alfonso d'Este, c.1523–1534

the central panel shows the forge of Vulcan, the "divine artificer" who wrought weapons for the gods and heroes of antiquity. The paintings for the chambers were also evocations of the classical world and for them Alfonso commissioned the leading artists of the day, including the old master Giovanni Bellini and Titian.

The alabaster chambers must have been the supreme achievement of 16th Century decor, but, unfortunately, no one will ever know how they looked. Even before the end of the century, with the death of the last Este, the works of art were dispersed; less than 50 years later the rooms themselves were destroyed by fire.

89

The paintings that Titian made for one of Alfonso d'Este's alabaster chambers take the ancient myths of love and physical abandon and free them from the cold, disciplined casing of the classical style: Titian's pictures pulse with ecstatic life. The bacchanalian scene at the left, below shows the island of Andros. There, according to the Greek philosopher Philostratus, a spring flowed with pure wine, and so the islanders lived in a state of happy, perpetual intoxication. Fittingly, Titian inscribed the picture's message on a sheet of music in the center foreground: "He who drinks and does not drink again, does not know what drinking is." Here, the artist fully indulged the sensuality of his patron, for whom he had once retouched the dresses on a painting by Bellini to make them more revealing. But Titian, even when complying with instructions, was never lascivious; there is a sense of courtly restraint and a feeling of contemporary reality that lend the scene dignity; amid the pleasures of drunkenness— represented by the paunchy man guzzling wine at the left, the sprawling drunken giant in the far right background and the sated, luxuriating nude in the right foreground— are the graceful ladies in the center of the picture. Dressed

Bacchanal of the Andrians, 1518-1519

with careless coquetry and surrounded by an aura of music, they could be aristocratic members of Alfonso's own courtly social set at a pastoral feast, drinking, dancing and flirting quite freely in the open air.

Love is also the theme of Titian's *Worship of Venus* in which a statue of the goddess of love and fertility watches over a frolicking field of cupids. Titian has communicated a sense of ripeness, of blooming, lusty physicality combined with an innocent playfulness. Again, as in his *Bacchanal,* Titian has evoked a mood by uniting myth and reality. Here love's gay emissaries—shown in detail on the following pages—like energetic kindergarten tots, fly about the branches of ripe apple trees, plucking fruits to drop to their fellows on the ground; they kiss and sweetly fondle one another; a mischievous imp takes aim with a bow and arrow, Cupid's familiar weapon. At the right side of the picture, strikingly real and womanly nymphs offer the marble goddess the gift of a mirror—the traditional symbol of Venus. Perhaps it is meant to bring her to life in reflection. All the way into the background of the scene, Titian has created a world of flesh—acres and acres of rosy, pinchable flesh—engaged in a charming, dimpled orgy.

Worship of Venus, 1518-1519

Portrait of Isabella d'Este, 1534–1536

Portrait of Federigo Gonzaga, c. 1525–1528

Isabella d'Este *(above left),* Alfonso's older sister, was of a different generation from Titian's principal patrons, but she shared their enthusiasm for his work. In her sixties she asked Titian to make the portrait of her shown here, looking as she did in her twenties. He based it on one that had been painted years before. Vain or nostalgic she may have been, but Isabella was also a most accomplished and ambitious woman. She proved herself a capable manager of state affairs and a skillful diplomat; she was extremely well read, a quick and witty conversationalist, and even sang and played the lute. At the age of 15 she had married Francesco Gonzaga, Marquis of Mantua, and her avid patronage of art made their court, although it was less wealthy and more provincial than others, a rival in splendor to that of the duchy of Ferrara. She felt that the high point in her career was when the Emperor, recognizing her loyalty, named

her son Federigo Gonzaga *(above right)* Duke of Mantua.

Federigo quite naturally inherited a love of art. He commissioned a number of religious paintings from Titian, in whose work he found spiritual reverence conveyed with a deep sense of humanity. In a letter concerning one picture, Federigo begged the artist to make it as "beautiful and tearful as possible." The quality of physical and emotional beauty that Federigo desired, Titian brilliantly provided in the moving and traditional drama of *The Entombment.* Here, the fact of Christ's death is made vivid not only by the ghastly pallor of His body, but even more poignantly by the grieving expressions of the mourners. In the *Madonna with the Rabbit,* the spiritual quality is heightened by its temporality—Titian has painted the scene as a man who has known what it is like to be a father and who has observed the tenderness of a mother with her child.

MUSÉE DU LOUVRE, PARIS

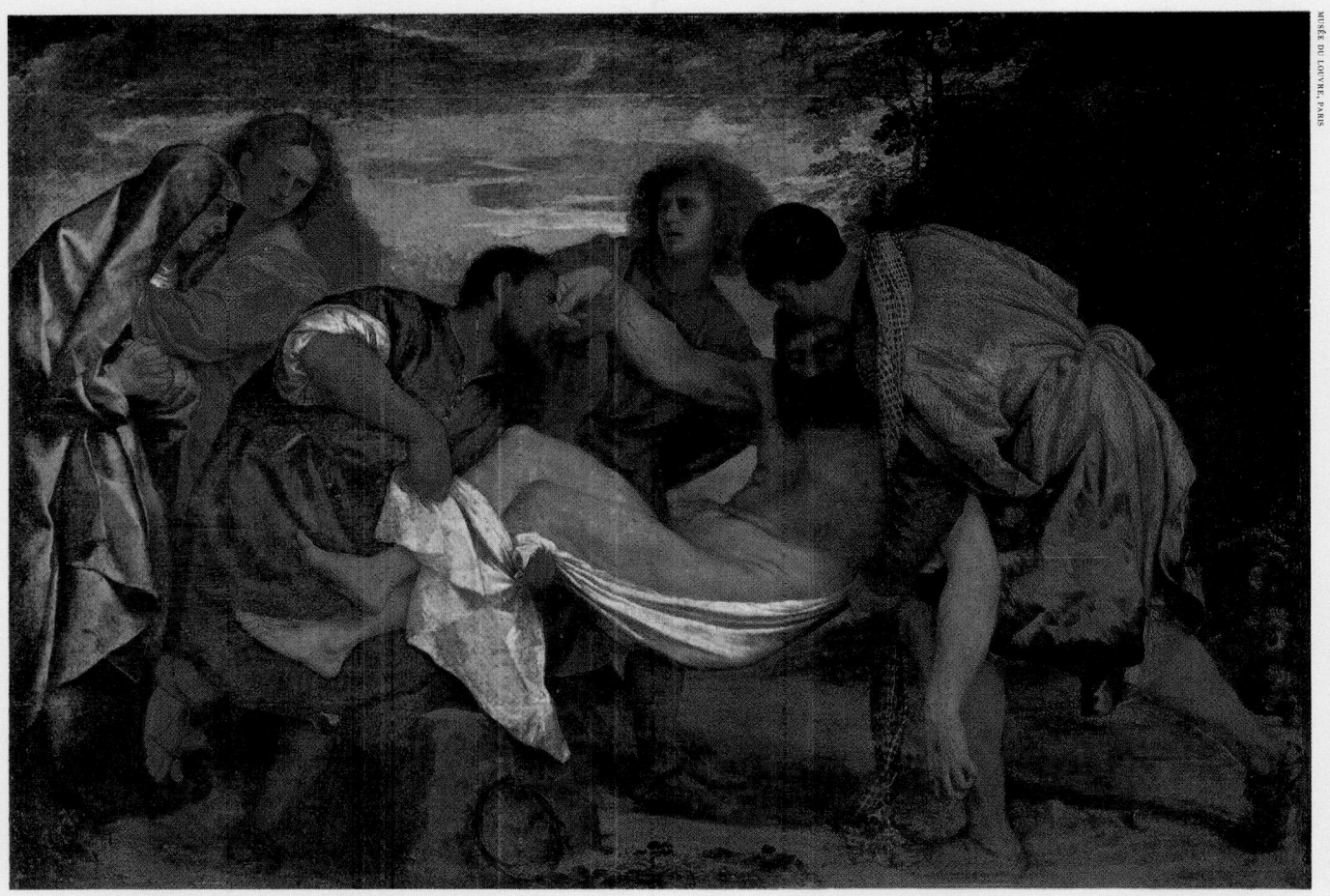

The Entombment, 1525

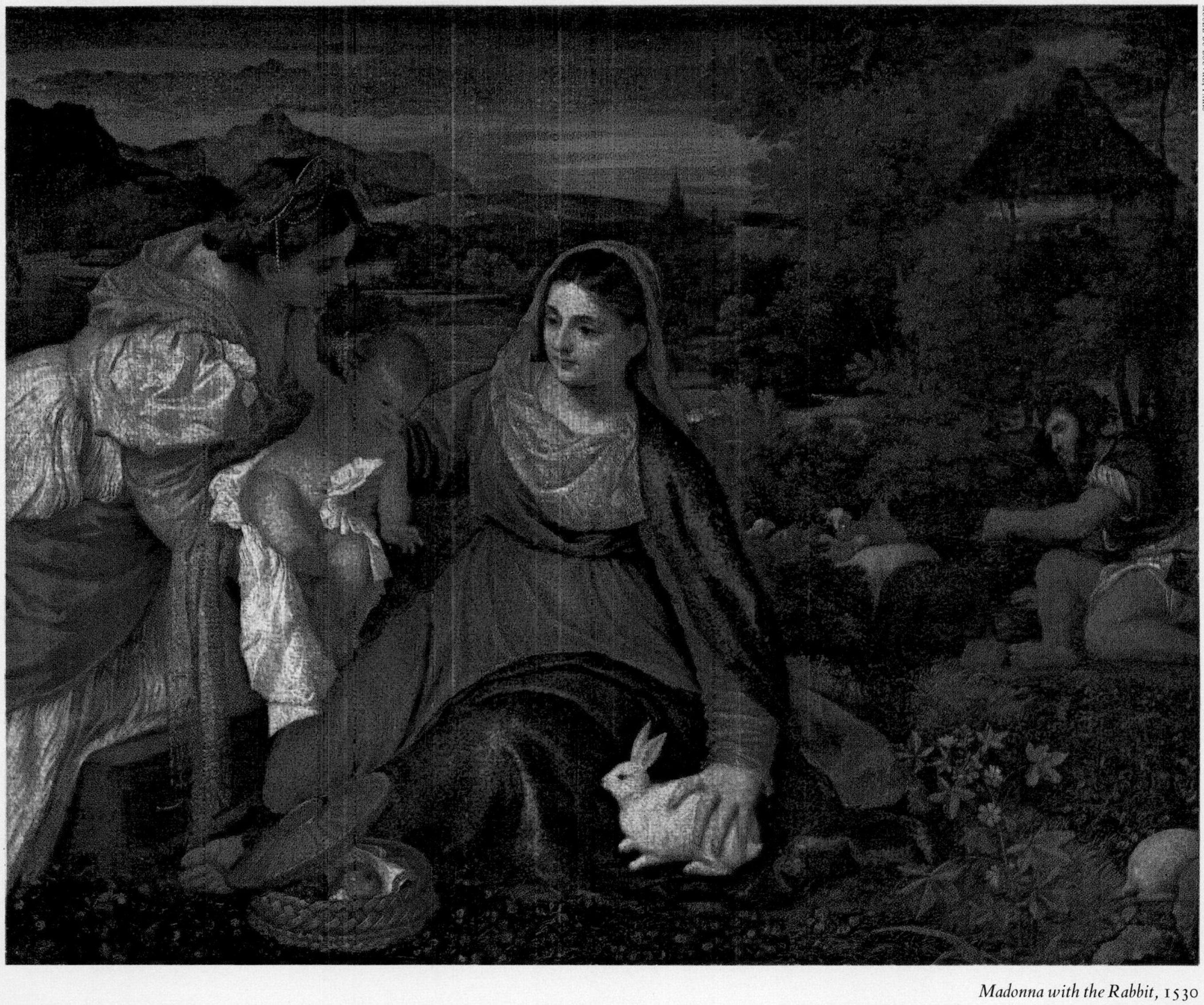

MUSÉE DU LOUVRE, PARIS

Madonna with the Rabbit, 1530

Portrait of Francesco Maria della Rovere, 1536-1538   *Portrait of Eleonora della Rovere, c.1538*

It was more than flattery that led the Renaissance writer and poet, Baldassare Castiglione, to find "wisdom, grace, beauty, intelligence, discreet manners, humanity, and every other gentle quality" in Eleonora Gonzaga *(above right)*. Castiglione had been a resident in the court at Urbino for five years when Eleonora, the daughter of Isabella d'Este, arrived to marry Francesco Maria della Rovere *(above left)*, the new Duke. But neither the poet nor the lady had need of fawning compliments. Rather, Castiglione was articulating his notion of the perfect individual, the man or woman in whom, by birth, by education and by design, the finest qualities were represented. His great work, *The Book of the Courtier,* was written about the court at Urbino under the rule of Guidobaldo da Montefeltro, Francesco's uncle. In it Castiglione set forth a kind of code of ideal behavior, summarizing a widespread humanist belief in the perfectibility of man and holding up beauty and love as central cultural values. It was a wonderfully enthusiastic

credo for developing the Self as a work of art. And, although the book sounds somewhat naïve, romantic and supercilious in the 20th Century, to early 16th Century readers of the favored class it was a manual for living, including hints on dress, cosmetics, speech and demeanor.

In Castiglione's view, grace was the prime quality, and, while he asserted that grace was basically a gift of nature, he insisted that it could be developed by instruction. Beauty, the natural result of grace in all things, was a primary attribute of women, and nowhere is this ideal better expressed than in Titian's paintings. His *La Bella,* for example, commissioned by Francesco della Rovere and probably depicting a courtesan at Urbino, is grace personified: her hair is carefully, yet simply, done; her jewelry is restrained but elegant, and her gestures are poised yet relaxed. She is a vision for those sensitive courtiers whom Castiglione describes as "so enraptured when they contemplate a woman's beauty that they believe themselves to be in paradise."

96

La Bella, c.1536

97

Titian's best-known and perhaps most perfect image of feminine beauty is his *Venus of Urbino.* It is especially fitting that the picture should have been painted for Guidobaldo della Rovere, the son of Francesco della Rovere and Eleonora Gonzaga, at whose court Castiglione had gathered the material for his great work.

The model for the painting, whose face is that of *La Bella,* may have been Guidobaldo's mistress. But what is important about the picture is that it is not simply a portrait of an individual, but of *all* beautiful women. Titian painted what Castiglione described: beauty is "that which is seen in the human person and especially in the face, and which prompts ardent desire. . . infuses itself therein and shines forth . . . with grace and a wondrous splendor . . . like a sunbeam striking a beautiful vase of polished gold set with precious gems."

Titian's Venus is of the Renaissance, not of the past; neither a marble figure in a garden nor an unreachable bacchante, she is a living being. Resting after a bath and attended by her maids, Titian's Venus lives in an Italian *palazzo;* her hair is in slight disarray, her flesh is toned with the blush of health, her gaze is frank. The bouquet she holds so casually, the easy modesty of her pose, the demure luxury of her single bracelet—these are the signs of her grace, the marks of her perfection.

Titian's painting speaks for itself not only as the product of a period's taste and an artist's skill, but also as a timeless object of sheer visual delight.

Venus of Urbino, c.1538

V

The
Triumvirate

At the end of June 1519 an election took place in the Holy Roman Empire that was to have profound consequences both for Italy's future and for Titian's future as a painter. When news of the election reached Venice, Titian was busily engaged on the altarpiece for the Pesaro family and on various projects for the Duke of Ferrara, but undoubtedly he took time off to gossip about the event with his fellow artists. Although he was never deeply concerned with politics or public affairs, except as they affected his own work, the naming of a new Emperor was a momentous affair. In January Maximilian I had died, and in June, after much negotiating and bargaining among the ruling families of Europe, Maximilian's 19-year-old grandson Charles was chosen to succeed him as leader of the Holy Roman Empire. He was to become Titian's greatest patron and lifelong admirer as well.

The Empire was an entity in little more than name, but still it was a name to conjure with. Although the German states within its borders were fractious and disunited, and the states it claimed in northern Italy were also claimed by other princes, the Empire potentially had vast political power, and the widespread awareness of this potential lent authority to the man who wore the crown. When Maximilian died there had been two major rivals for his throne: Charles and Francis I, King of France. Charles was a pale, awkward young man who was often ill. He had inherited the physical attributes for which many of his Habsburg forebears were noted: a jutting jaw, a prominent nose and a swollen underlip, which gave him an adenoidal appearance. But in personality, Charles resembled his grandfather only in his passion for hunting. Maximilian, whom history has dubbed "the last of the knights," was flamboyant and headstrong, an extrovert swift to battle and addicted to chivalric display. Charles was quiet and undemonstrative and fond of family life. He liked to read and he appreciated art, although he was never the model of the cultivated Renaissance gentleman. He was, however, instinctively adroit at statecraft, a useful skill for a man whose sprawling family domains included the Netherlands, Austria, Spain and the Spanish possessions in the New World, Naples, Sicily, Sardinia and parts of France. Very early he

Titian caught the character and spirit of his good friend Pietro Aretino with such precision that the writer once declared of his portrait: "Though I be painted I both speak and hear." The picture was made when Aretino was 53, and it shows the lusty ex-court jester full of brawling health.

Portrait of Pietro Aretino, 1545

showed a conscientious regard for the problems connected with holding his enormous inheritance together.

His rival in the struggle for the Imperial crown, Francis I of France, was six years Charles's senior and as unlike Charles as he could be—daring and reckless, given to gaiety and show, excessively extravagant. At 14 Francis had appeared at a tournament one day dressed from head to foot in cloth-of-gold and the next day similarly garbed in white satin. At 19 he already had a painter and a sculptor in his personal employ. Eventually his love of the arts was to bring Leonardo da Vinci to live in France, and he was to elicit from Benvenuto Cellini ecstatic praise as "that glorious monarch." The esteem was more or less mutual; Francis is said to have remarked with kingly generosity: "I can create a peer but only God can create an artist."

Tall, handsome, broad-shouldered and big-chested, Francis had only one physical defect—a slight squint from near-sightedness. A Venetian envoy to France reported that he was "a man of inexhaustible endurance . . . ever at the chase—now after a stag, and then after a woman." Yet this dazzling figure proved no match for his younger rival in their contention for the Holy Roman Empire. Both Charles and Francis saw the title of Emperor as conferring not land so much as nominal leadership of the Christian world. But to Francis the title was a convenient fiction to further his other ambitions, while Charles regarded it as a family asset, an important link in the continuity of Habsburg power.

To become Emperor required the votes of seven princely Electors, and to gain those votes Charles and Francis conducted a stiff competition in bribe-giving. Francis, however, lacked the necessary cash and could not capitalize on his credit, while Charles had the backing of the powerful German banking house of the Fugger family. This Renaissance institution, built upon commerce, mines and land, underwrote Charles's candidacy to the tune of half a million florins. It was a debt of which he was never to be wholly free, but it did buy his unanimous election. In 1520 Pope Leo X recognized Charles as Emperor, although a decade more was to elapse before the Imperial crown was officially placed on Charles's head.

There were several reasons for the delay. Almost immediately after the election Charles became involved in a war with Francis. He also faced troubles within his own huge realm. In Spain there were rumblings of revolt as nobles and townsmen, jealous of their local autonomy, resisted the idea of rule by an absentee monarch whose policies toward Spain were dictated by foreign advisers. In Germany Martin Luther and his followers were challenging the established Church, and peasant uprisings turned many parts of the land into scenes of bloody riot. In Austria Ottoman armies encamped along the lower reaches of the Danube cast an ominous shadow over the eastern end of the Empire. Over all, however, loomed the threat of a French king angered by the results of the Imperial election. In 1521 Francis declared war on Charles. His immediate objective was to strengthen his hold upon the duchy of Milan, long disputed by France and the Empire. Francis's predecessors, Charles VIII and Louis XII, had both crossed the Alps to enforce their Italian claims, and Francis himself

This bizarre composite monster was one of the illustrations in a bitter anti-papal pamphlet written by Martin Luther and his colleague Philip Melancthon. Alleging that the beast was found in Rome at the end of the 15th Century, they analyzed it as a symbol for the papacy: the ass's head, inappropriate to the beast's body, was just as incongruous as the Pope at the head of the Church. The left foot was clawlike, they charged, because the Pope's servants grasped the world so greedily; the right hand was an elephant's foot because the overbearing Pope crushed weak souls with elephantine force. The old man's head facing backward meant that the papacy was drawing near its end, and the dragon-headed tail spewed fire just as the Pope "belched" edicts, books and words. The beast's scales, they asserted, represented reactionary rulers clinging to the Pope.

had previously made the trip: soon after his coronation in 1515 he had swept over the mountains in a brilliant five-day march to seize the city of Milan. At the time Francis was only 21, and in the heady aftermath of battle he ordered himself knighted by the Chevalier Bayard, the noblest soldier in his realm. The chevalier was known far and wide as the epitome of the ideals of medieval chivalry, and was called the "knight without fear and without reproach." Yielding to Francis's whim, he wryly observed that this was "the first king I have ever knighted."

This time, however, it was the Imperial forces that were successful at Milan. Charles's army retook the city from the French in November 1521 and then went on to capture the nearby towns of La Bicocca and Robecco. It was in the latter battle that the incomparable Bayard lost his life, victim of a ball from the precursor of the musket, the weapon that was to end forever the medieval concept of warfare. Around the same time Francis also lost the services of his commander-in-chief, Charles de Bourbon. Insulted and slighted by his King, Charles was induced to change sides, and he ended up by capturing Francis at the battle of Pavia in northern Italy in 1525.

Before long the royal prisoner was freed on condition that he sign a treaty of peace—a pact he blithely broke five months later by joining the so-called League of Cognac, which allied the Pope, Venice and France against the Imperial armies. This maneuver so enraged Charles that he challenged Francis to a duel. Francis accepted, but the contest never took place; in the final analysis neither man was willing to risk his kingdom on the outcome. And so the war dragged on, the tides of battle rising and subsiding, favoring first one side and then the other, like some nightmarish chess game in which there are endless checks but no checkmates. During its course, in 1527, Imperial forces sacked Rome; over a horrendous nine-month period of looting and debauchery, the churches and palaces of the Eternal City fell into smoking ruins; the Sistine Chapel stabled the horses of the invaders. Barricaded in his citadel in the Castel Sant' Angelo, the Pope cried, "Why died I not from the womb? Why did I not give up the ghost when I came out of the belly?"

By a curious turn of fate, it was the sack of Rome that brought together Titian and two other men in an inseparable lifelong friendship. One was the sculptor and architect Jacopo Sansovino; the other was the famous and infamous Pietro Aretino.

Aretino, a poet and playwright who was also the world's first gossip columnist, was to play a vital role in Titian's career. A shoemaker's son, born the year Columbus made his first voyage to America, Aretino had reportedly been forced to leave his native town of Arezzo at 13 because he had angered the local clergy by a scurrilous sonnet about Church indulgences. Like many another gifted young man, he made his way to Rome and there found haven in the house of a wealthy banker and patron of the arts, Agostino Chigi. Aretino's esthetic tastes were formed by the writers and painters who were Chigi's frequent guests, and he may also have picked up from Chigi the blend of canniness and extravagance that characterized his behavior ever after. Many stories were told about Chigi's grandiose gestures. One concerned a lavish banquet he gave at which the servants cleared the table by throwing the gold plates out the window into the Tiber River—where they were caught in a net and quietly hauled in after the guests had gone.

In time Leo X heard about Aretino and invited him to the Vatican as a kind of court jester. Gradually he acquired a reputation as a writer of letters and satiric pieces—called pasquinades—on the political issues of the day. He also won acclaim as a poet of some charm, turning out delicately lascivious verses and flattering sonnets in praise of possible patrons. Then, in the summer of 1524, Aretino found himself in trouble. A scandal erupted over some lewd verses he wrote to accompany a group of lewd drawings by Giulio Romano, a former pupil of Raphael, and he was forced to flee Rome. For the next few months he wandered restlessly from town to town; for a time he attached himself to the entourage of an old Roman friend, General Giovanni de' Medici, better known as Giovanni delle Bande Nere—Giovanni of the Black Bands—for the black trappings worn by the papal troops he commanded.

By November, however, Aretino was back in good graces in Rome, composing flattering odes in praise of a new Pope, Clement VII. "He now walks through Rome dressed like a Duke," reported Francesco Berni, a Vatican secretary. "He pays his way with insults couched in decorative words. He talks well and he knows all the scandal of the city. He hobnobs with the Este and Gonzaga families. Them he treats with respect, others with arrogance. He lives on what the former give him. He is feared for his satiric wit and enjoys hearing people call him a cynical, impudent slanderer. The Pope has bestowed on him a fixed pension as the reward for having dedicated certain mediocre verses to His Holiness." Indeed, Aretino's "satiric wit" endeared him to some and made others his deadly enemies. The Marquis of Mantua referred to him as "one of my cordial friends," but the austere Giovanmatteo Giberti, an important Vatican official who disapproved of Aretino and was a target for some of his sharpest barbs, was said to have instigated an attempt on the poet's life. On his way home late one evening, Aretino was set upon by assailants and stabbed. He recovered, but at the cost of a permanently crippled right hand, which forced him to write thereafter by holding his pen between his thumb and two last fingers.

In 1526, possibly bored with Roman high life, Aretino again joined the camp of Giovanni delle Bande Nere. The general was heading northeast to battle an Imperial army sweeping down into the Po Valley, the same army that was soon to engulf and sack Rome. In an engagement just outside Mantua Giovanni was mortally wounded. He was hit in the thigh by a cannon ball, and his gangrened leg had to be severed. Aretino movingly recorded his hero's final hours:

> [The doctors] asked for eight or ten persons to hold him while the agony of the sawing lasted. "Not even twenty," he said with a smile, "would be able to hold me." With a perfectly calm face he took the candle in his own hand to give light to the doctors. Whereupon I fled and, stopping my ears, I heard two groans only, and then I heard him calling me. When I came to him, he said, "I am cured!" and turning himself this way and that, he made a great rejoicing. And if the Duke of Urbino had not

restrained him, he would have had us bring him his foot, with the piece of leg still clinging to it, laughing at us because we could not bear the sight of what he had suffered. . . .

The pain, which had left him for a while, returned two hours before dawn with every kind of torment. Hearing him beat upon the wall in a frenzy, I was stabbed to the heart, and dressing in a moment, I ran to him . . . he begged me to read him to sleep. . . . Finally, having slept, it may have been a quarter of an hour, he awoke and said, "I thought that I was making my will and here I am cured. . . . If I keep on getting better like this, I'll show the Germans how to fight and how I revenge myself." Then the light failed and he asked for extreme unction. He received the sacrament and said, "I don't want to die in all these bandages." And so we brought a camp bed and placed him on it, and while his mind slept, he was taken by death.

Aretino did not return to Rome after Giovanni died because he suspected that the city would fall—and, in fact, so predicted. Instead he went to Venice, there to live for the rest of his life. "The diadem of the world," he called it, more in true love than flattery. "All other places seem like furnaces, hovels and caverns in comparison with my most beloved, most excellent and adorable Venice." By 1529 he had leased a large house overlooking the Grand Canal, which he proceeded to furnish with gifts from great and near-great personages of the day—wall hangings, cabinets, desks, beds and chairs. Some of these gifts were genuine, but many were thinly disguised bribes, given to Aretino to shut him up when he came into possession of intimate bits of gossip. In time the splendor of the poet's house rivaled that of the homes of his patrons. A handsome bust of the building's owner decorated its front entrance, and in its glass-roofed reception hall there was a conspicuous display of his correspondence with noblemen and ladies, princes of the Church, wealthy merchants and famous artists. In these sumptuous surroundings, and with the services of a good cook, Aretino settled down to enjoy life and play host to his friends. One of the most intimate of these was Titian. Another was Sansovino, who, like Aretino, had come to Venice from Rome around the same time and for similar reasons.

As a child in Florence Sansovino had been secretly taught to draw by an indulgent mother who hoped he would grow up to be as famous as another Florentine whose star was then rising, Michelangelo. And indeed, many years later in Rome, Sansovino's sculptures were thought to be superior in some respects to Michelangelo's. "His marble draperies were slender," wrote Vasari, "well executed in beautiful folds, showing the lines of the body, while he made his children soft, tender, without such muscles as adults have. . . . His women are sweet and charming, and of the utmost grace. . . ." The artist's nature must have been similarly sweet. Vasari notes that "he did not show much diligence," but that he made up for it in talent: "He produced his work with ease . . . and a certain lightness pleasing to the eye." Also, although he was a redhead, with a redhead's quick temper, "it was soon over, and often a few humble words would bring tears to his eyes."

In Rome Sansovino was prized as an architect as well as a sculptor. His design for the Church of St. John the Baptist was picked in competition over the entry of no less an artist than Raphael, and his palaces, loggias and triumphal arches were sought after not only by noblemen but also by cardinals and no fewer than three popes. "All Rome was in his hands," writes Vasari. But in May 1527, when the Imperial army began its sack of Rome, Sansovino fled. Unlike Benvenuto Cellini, who was a crack shot and who claimed to have picked off the commander of the Imperial forces as they breached the walls, Sansovino was a man of peace.

He headed north, intending to seek the favor of the King of France; by now Francis's enthusiastic patronage of creative people was common knowledge. But en route Sansovino stopped in Venice to replace the clothing and other necessities he had left behind in his flight. The Doge, hearing of his presence and knowing of his reputation as an architect, invited him to stay. The centuries-old Basilica of St. Mark was in desperate need of repairs. Its foundations had been undermined by water, and its domes were dangerously cracked. The Doge asked Sansovino to tackle the problem, and Sansovino consented. He reinforced the structure of St. Mark's with massive beams, rebuilt its rotting piers and repaired its damaged walls. In gratitude, the Venetian senate gave him the post of Protomaster of the Procurators of St. Mark's, the equivalent of chief architect of the republic, and granted him a house and salary befitting the station.

Sansovino, Aretino and Titian became boon companions. Venetians soon dubbed them the Triumvirate, and apparently they became something of an institution in fashionable circles. They entertained one another at little dinners graced by good conversation, handsome women and tasty viands sent by admirers—to whom Aretino never failed to pen such elegant letters of thanks as this one: "Most kind, most dear, and most gracious Messer Niccolò! Because it seemed to Titian, who gives life to colors, and to Sansovino, who gives breath to marble, an almost ungrateful action by thanking you alone for the gift of pickled fennel and spice cakes, they both with me and with the testimony of their appetite . . . confess to being much obliged to you." Or: "For the fine and excellent turkey which the affable kindness of your true courtesy sent me from Padua, I give you as many thanks as he had feathers. . . ."

It must have been a happy relationship. The three men loved life with equal gusto, and they had a healthy respect for each other's talents. Titian painted Aretino's portrait several times *(page 100)*; Sansovino included the faces of his two friends—as well as his own—on the bronze doors he designed for St. Mark's sacristy; Aretino praised Titian and Sansovino ceaselessly and advanced their affairs in a flood of letters to every famous person of the time. These letters, generously laced with flattery, boasting and malice, have been used by scholars to attack Aretino. They have labeled him a cynic, a depraved calumniator, a parasite who battened on the corruption of others like a "poisonous fungus on a dunghill." But the artists who were Aretino's friends applauded him. When he lashed out at princes and popes his words struck a blow in their behalf; many a painter or poet spent his free time trying to collect money owed him by lordly patrons. So these creative people openly approved when the poet Ariosto

Elevated shoes—such as the ones shown below—helped Venetian ladies reach new heights of fashion during Titian's time. Called chopines, they were worn throughout western Europe (probably having been introduced from some place in the East) but were especially favored in Venice as an aid in navigating the frequently flooded or muddy piazzas. There were some cynics who said that chopines were designed to keep Venetian women at home, for they sometimes were 18 inches high, and special attendants had to support the ladies as they took each precarious step.

characterized him as "the divine Aretino, the scourge of princes."

Aretino has been underestimated as a writer. For centuries his plays and *Dialogues,* along with his letters, have been regarded as remarkable mainly for their obscenity and have been sold under the counter as pornography. In fact, they are often biting commentaries on the manners and mores of his day, full of sharp observations and pungent humor. His whores and hypocrites, greedy merchants, two-faced courtiers and bogus pedants were real and recognizable. There is irony as well as comedy in the advice given a mother in one of the *Dialogues* to make her daughter a prostitute: "For the nun betrays the sacrament, and the married woman assassinates the sanctity of matrimony, but the whore . . . like a soldier . . . is paid for doing wrong, and hence she is not to be criticized. . . ." Prudish readers have turned away from Aretino in disgust, but his delighted contemporaries thought him one of the keenest wits and best writers of the day.

Aretino's writings, moreover, tell us much about Titian's Venice; they open a window onto the Grand Canal and the pleasant world of the Triumvirate. He notes the passing barges piled high with melons "as if forming an island," the shops full of game and vegetables, the vendors crying their wares. He rhapsodizes over the charms of the Venetian women, blonde beauties famed throughout the world: pretty young housewives shopping in the market as well as ladies of fashion in silk and gold, mincing along in high wooden clogs as though walking on stilts. He describes cozy dinners and hectic times when his house is so crowded with out-of-town guests that he has to slip away to friends' homes for a

A cut-away view of a Venetian woman's costume shows her lofty chopines *(right)* and the long gown that normally covered them *(left).* Speaking of women wearing such shoes, an English visitor said derisively that "Venice is much frequented by May-poles." Another remarked that Venetian ladies were composed of three things: wood, clothes and woman. A reference to these shoes even found its way into Shakespeare, whose Hamlet sarcastically greets a woman with, "Your ladyship is nearer heaven than when I saw you last, by the altitude of a chopine."

bit of peace and quiet. He speaks of gondola races on the canals; of seeing a boatload of drunken Germans upset in the cold water; of one particularly splendid sunset in which the sky was full of great white clouds and the city seemed magical, the houses "like fairy palaces," deep shadows contrasting with dazzling marble and vermilion roof tiles. "Nature, mistress of all masters!" he cries. "How miraculous is her brush, how wonderful her pencil!"

Aretino, outgoing and ebullient, was probably the liveliest of the Triumvirate, but Sansovino must have run him a close second. Vasari says that the sculptor was "handsome and graceful, so that many ladies of rank fell in love with him," and notes discreetly that for a time he was the intermittent victim of "some disorder caused by the escapades of youth." Titian, of the three, was definitely the most reserved. "I marvel at him," wrote Aretino, "for no matter with whom he is or where he finds himself, he always maintains restraint. He will kiss a young woman, hold her in his lap and fondle her. But that's as far as it goes. He sets a good example for us all." There was a reason for this self-restraint. Titian, by the time the trio first got together, was no longer a bachelor. He had married in 1525, and the joy he took in his friends' companionship was tempered by domestic concerns. His wife, a barber's daughter from a small village of Cadore, was a young woman who had been his housekeeper and mistress for some five years. Cecilia had already borne Titian two fine sons, Pomponio and Orazio, when in 1525 she fell seriously ill. Titian, wishing to legitimize the children, married her. He must have done so out of love, for bastardy carried no particular stigma, and he could have legally recognized the boys as his sons without marrying their mother. In any case the wedding was a happy affair, which apparently revived Cecilia, for she recovered and bore Titian two more children, both daughters; only one of them, Lavinia, survived.

With the responsibilities of a growing family, Titian was kept busy not only with his work but also with trying to collect money owed him. He engaged in a protracted quarrel with the head of the brotherhood of St. John and St. Paul over the payment for the *St. Peter Martyr* altarpiece. He also had his money troubles with the Duke of Ferrara; in addition to the painting the Duke had ordered for his alabaster chambers, Titian was producing all sorts of other paintings for Alfonso. In a letter dated 1523, Titian complains to his patron of not having enough money for decent clothes in which to appear at the Duke's court, and reminds him that he has already given him three paintings, any one of which is worth 100 ducats, but that the Duke's agent has so far presented him with only 100 ducats for all three.

A more recent patron was also being difficult about money. This was Federigo Gonzaga, Marquis of Mantua, son of the famous Isabella d'Este and nephew of the Duke of Ferrara. Titian had come to Federigo's attention during his stopover in Mantua in 1523 en route to deliver the *Bacchus and Ariadne* to Ferrara. The young marquis had begged his uncle Alfonso for the loan of Titian's services, and around 1525 Titian painted a portrait of him, which now hangs in the Prado in Madrid *(page 94)*. It shows a young man with haughtily raised eyebrows, dressed in a gold-laced tunic, holding a gold court-sword in one hand and fondling a dog with the other. Some years earlier Titian had done a similar portrait of Federigo's uncle, but, although the Duke appears in the same pose *(page 89)*, there are amusing differences. Alfonso's shrewd face looks out over a plain dark doublet and coat, and one hand rests on a serviceable swept-hilt sword while the other rests on a cannon.

Many of the works that Titian produced for the marquis have been lost or cannot be identified. One of them is thought to be an *Entombment* that now hangs in the Louvre *(page 95)*. In this work the portrayal of Joseph of Arimathea, shown holding Christ at the knees, is believed by some scholars to be a portrait of Aretino. For one of the rooms of Federigo's palace at Mantua, Titian painted a series of portraits of the Twelve Caesars, using as his models antique medals and marble sculptures. The Caesars, Lodovico Dolce wrote, were "of such perfection that people go to Mantua to see them, thinking they see the Caesars themselves, not merely pictures." After Mantua's war with Austria in the 17th Century, the portraits were taken to England and then to Spain, where they disappeared; the only traces of them are some copies by other painters and a set of engravings by Egidius Sadeler, the 16th Century Flemish engraver.

Sometime in June 1527 Titian also sent his new young patron a portrait of Aretino painted shortly after the poet moved to Venice. With it went a letter describing Aretino as "a second St. Paul [come] to preach the virtues of your Excellency," and hinting to the marquis that he pay Titian what he thought the painting worth. Federigo replied with expressions of the deepest delight. "I . . . thank you, and shall hold these pictures dear for your sake; and you may be assured that nothing you could have done would have been more agreeable to me, or make me feel myself under more obligation. When I can I shall ever be ready to do you a pleasure. . . ." But Federigo's readiness to reward Titian was not precisely speedy; six months later the artist was still living on promises.

There must have been other times, however, when the marquis was more trustworthy, for Titian continued to send him paintings. In February 1530 one of Federigo's Venetian agents wrote to report that Titian was well advanced on a *Madonna and St. Catherine,* as well as on a portrait of the marquis himself, but that a picture of "bathing women" was still only in the design stage. The *Madonna and St. Catherine* is believed to be the painting now in the Louvre known as the *Madonna with the Rabbit (page 95),* a picture notable not only for its beauty but also for the possible identity of its models. Titian's wife Cecilia may have posed for the figure of the Madonna, while the bearded shepherd in the background may be Federigo.

In March 1530 Titian wrote the marquis to apologize for the lateness of the painting of the bathers, explaining that he had been troubled by a painful skin irritation. He also expressed thanks for a gift that the marquis had "condescended with habitual grace and liberality to make in my favor." The gift may have been a clerical benefice for Titian's firstborn, Pomponio, an appointment for which the artist had been angling for some time. His gratitude, however, was premature. A year later he was grumbling that although the boy had donned clerical garb and was being congratulated all over Venice for the marquis's bounty, the income from the benefice had not yet arrived. A year more was to elapse before the matter was settled.

Meanwhile tragedy struck Titian's household. In August 1530, while Venice steamed in the heat, Cecilia died. "Our master Titian is quite disconsolate," wrote Federigo's agent. "He told me that in the troubled time of her sickness he was unable to work on the portrait of the Lady Cornelia or at the picture of the nude that he is doing. . . ." Cecilia was buried on August 5th, and soon afterward Titian sent for his sister Orsa to take charge of his household. The following year he moved to a house in the pleasant district of the Biri Grande, looking northward over the lagoon. Here he was to live the rest of his life.

The flavor of those days was captured by Francesco Priscianese, a Latin scholar who was a guest at a dinner party at Titian's house:

> On the first of August I was invited to celebrate that bacchanalian feast called Ferr'agosto. I don't know why it is called that, although we had a long conversation about it in the delightful garden of Titian Vecelli, that most excellent painter who affably provides the best of entertainment. With him were assembled, as like goes to like, some of the rare geniuses at present in this city; first, Messer Pietro Aretino, a new miracle of nature, and next to him Messer Jacopo Tatti, called Il Sansovino, the great imitator of nature with a chisel, as our host was with the pencil. . . . Although the place was shaded, the sun still had heat, and here, before they spread the tables, the time passed in contemplating the living representations of the excellent paintings of which the house is full, and in talking about that truly lovely garden. . . . It is situated at the farthest part of Venice on the edge of the sea looking over to the lovely isle of Murano and other beautiful places. As soon as the sun sank, this part of the sea was covered by a thousand gondolas, adorned with the handsomest ladies and resounding with vocal and instrumental music, which still at midnight accompanied our delightful supper. . .which was in no way less excellent or worse ordered, than it was plentiful and well-furnished, besides the fine dishes and wines; and all the pleasures and enjoyment were in unison with the quality of the time, of the persons and of the entertainment.

In 1532, a year after he had moved to the house in the Biri Grande, Titian acquired the most impressive of all his patrons: the Emperor Charles V. The war between Charles and Francis had ended in 1529. In November of that year, Pope Clement VII, as spokesman for the League of Cognac, had journeyed north to Bologna to meet Charles, and there the two leading princes of Christendom signed a treaty of peace. On February 24, 1530, Clement finally placed the Imperial crown on the head of the German monarch, giving him official Church recognition as Holy Roman Emperor. That same day Charles became 30 years old. If Clement performed the ceremony reluctantly he gave no sign of it; Bologna—one of the papal states—turned Charles's coronation into a typically Renaissance festival. The streets were hung with banners and flags; the Cathedral glittered like a giant jewel box. Clement and his cardinals in vestments of cloth-of-gold, and Charles in gilded armor and an Imperial mantle of gold brocade, played out their roles against a backdrop that included the purple robes of scholars, the scarlet and blue uniforms of the Pope's Swiss Guards, and the brilliant silk and velvet court dress of foreign ambassadors.

Affairs of state took Charles back to Germany almost immediately after the ceremony, but in the fall of 1532 he returned to Italy. On his way south he was received at Mantua by Federigo, who had prudently decided to attach himself to the "rising sun" of the Habsburg Empire, and who entertained the monarch royally. Charles must have enjoyed Mantua, for he loved fine armor and fine paintings and the palace had plenty of both. The Mantuan armory contained the armor of every Gonzaga who had ever ruled Mantua, and the art collection contained, of course, a number of paintings by Titian. Charles was especially impressed by Titian's portrait of Federigo, and the marquis was no doubt full of praise for his painter. The upshot was that in November 1532, after the Emperor had gone on to Bologna, Titian was called there to do a preliminary study for a royal portrait. The painting that resulted (*page 120*) is based partly on the study and partly on a portrait of Charles by the Austrian painter Jakob Seisenegger. But where the latter presents what the historian Hans Tietze calls "a pedantic inventory of all the characteristics of the Emperor's countenance and costume," Titian's painting is a living likeness, and a great work of art.

Just before Charles sat for the preliminary sketch for this portrait, a friend of Titian's, the sculptor Alfonso Lombardi, came to him and asked if he might be allowed to accompany him so that he might see the celebrated monarch at close range. Titian agreed, but once inside the room Lombardi did more than look. Standing behind Titian, he took out a little box of plaster and modeled a head of Charles while Titian worked at his sketch. When the sitting was over, the Emperor asked Lombardi if he might see what he had been doing, and was so pleased with the head that he asked the sculptor if he thought he could reproduce it in marble. Lombardi replied that he was sure he could, and went off happily to begin the task. The result, says Vasari, was "a most admirable work," but it was paid for at Titian's expense. When Titian finished the Emperor's portrait, Charles gave him the handsome sum of 1,000 *scudi* but instructed him to give half that amount to Lombardi. "Whether Titian felt aggrieved or not, we may all imagine," Vasari comments.

Beyond this incident, however, Titian had no reason to complain of the Emperor's generosity. Charles was delighted with his portrait and knew that he had encountered genius. From Barcelona, where he went in the spring of 1533, he sent letters of patent, creating Titian Count Palatine and Knight of the Golden Spur, titles that granted the painter entrance to the court and gave his sons the rank of nobility. And although Charles was no classicist, he gracefully noted that he was following an example set by Alexander the Great. Just as Alexander had elected to be painted by none but the great Apelles, so he, Charles, would henceforth be painted by none but Titian, the Apelles of his own day. "And so much was the invincible Emperor pleased with the manner of Titian," records Vasari, "that once he had been portrayed by him, he would never permit himself to be painted by any other person."

Image plus Insight

Titian was the most sought-after portrait painter of his day. In his journeys to paint members of ruling families—Habsburg, Gonzaga, Farnese and Medici—he traveled more than most other artists of the time—as far south as Rome and north through the Alps to Augsburg. So desirable were his portraits of the great and the mighty that he was asked to paint both Francis I of France and Turkish Sultan Suleiman the Magnificent, although his patrons knew that he had never laid eyes on either man, and existing portraits had to serve as his models.

In an art-conscious 16th Century Europe, it is not surprising that the master painter of the Venetians should have gained such an international reputation. But Titian succeeded not merely because he was in vogue, but also because his patrons prized the new freshness, warmth and vitality that he brought to portraiture. A keen observer of men and an extremely bold craftsman, Titian was able to discern and reveal those gestures and expressions that suggested nobility and elegance of character far more subtly than just costume or setting ever could. With a rare sense of drama, Titian probed personality and exposed it on canvas with directness and spontaneity. Although the modern viewer is some four centuries removed from Titian's subjects, the artist's evocative powers are so great that each individual comes alive with stunning impact. What is more, they appear in pictures that are invariably beautiful to look at.

Nonchalantly resting his arm on a window ledge or parapet, this cool, self-confident young man—who may be Titian himself—gazes obliquely at the viewer. It seems almost as if he might lean out to speak a word or two. Titian may have learned the compositional device of a window within the picture frame from Giorgione, and he cleverly used it to heighten the illusion of depth in the painting and to create a sense of intimacy.

Portrait of a Man (Ariosto), c. 1508-1512

The Pesaro Madonna, 1519-1526

Votive Portrait of the Vendramin Family, 1547

Portraits of wealthy benefactors had been included in religious paintings long before Titian's time. In the altarpiece at left, commissioned by Jacopo Pesaro, the donor and his family are shown in profile in the lower corners of the picture. Titian respected convention by keeping the portraits subordinate in the scene, but he invented a wholly new composition for the altarpiece. Earlier Madonnas, such as Giorgione's on page 65, showed Mary high in the exact center of the picture—a strong but static design. Titian moved the Madonna and Child to the right, creating an asymmetrical composition that draws the viewer's eye back and forth across the scene. The artist enlivens it even further with realistic interactions of glance and gestures between Mary and her attendant saints.

Using a similar composition in a painting *(above)* where a precious relic of the True Cross is the off-center focus of attention, Titian was able to go further in creating fully fleshed portraits of his patrons. Surrounded by his family, the patriarch of the Vendramin family kneels reverently, one hand resting intimately on the altar; beside him stands his brother Andrea, whose famous namesake once miraculously rescued the holy relic when it fell into a Venetian canal; at the far left is Andrea's eldest son, and in the corners are his six younger children, one of whom charmingly holds a puppy. Although devout, the scene is filled with the vigor of life.

Pope Paul III and His Grandsons Alessandro and Ottavio Farnese, 1546

Among the 200 or so formal portraits by Titian are a few that stand out as landmarks of the form. Most of his paintings are of single figures, sensitive character studies such as those on this page. One of them, the young boy *(below, left),* has a wistful expression, which betrays how ill at ease he is in the pompous trappings befitting a member of one of the great Italian families. But Titian also injected a new note into portraiture. His study of Pope Paul III and his grandsons *(left)* has been called "the most dramatic portrait yet painted." At center is Pope Paul, a shrewd 77-year-old who was both a great advocate of reform and also one of the most lavish dispensers of papal favors in the history of the Church. Behind him stands the pious Alessandro, whom the Pope had made a cardinal at 14. At right, Titian shows Ottavio in a genuflecting posture that suggests not humility but a fawning plea for privilege. History seems to bear out Titian's reading of the duplicitous character of Ottavio, who subsequently rebelled against the old Pope in an international contest over some of his father's territory. On his deathbed, the Pope repented his excessive nepotism.

Portrait of Cardinal Pietro Bembo, 1540

Portrait of Ranuccio Farnese, 1542

Portrait of Cardinal Ippolito de' Medici, 1533

The 16th Century artist-historian Vasari accurately described Titian's late paintings as "done roughly in an impressionistic manner, with bold strokes . . . ." In them, Titian eliminated much specific detail; here, for example, the cut and texture of the subject's coat are barely indicated, the background is nondescript. Compared with the precise quality of the Bellini portrait on page 24, this portrait provides much more than mere documentation. Whereas Bellini's Doge is stiff and wooden, Titian's young "Englishman"—his identity and true nationality are uncertain—seems alive and vital. Using the white ruffle and gold chain to focus attention on the young man's handsome face, Titian achieves an extraordinarily lifelike quality by setting off the warm flesh tones with deeper values of the same tones in the background, by posing the figure in a relaxed and spontaneous attitude, and by catching the subject's gaze as he directly, almost verbally, addresses the viewer. One senses that he has stopped only momentarily and may move again at any moment. It is as fine a portrait as Titian ever painted.

Portrait of a Man (The Young Englishman), c. 1540-1545

VI

A Gallery
of the Great

When Charles V received Titian's portrait of him *(page 120)* in 1533 and announced delightedly that henceforth he would be painted by no one else, Titian's reputation as a portraitist soared to new heights. Everyone who was anyone wanted to be painted by him, as a mark of status, and Titian was in the enviable position of being able to pick and choose from a multitude of would-be clients. "Knowing his own merit," wrote Lodovico Dolce, "he held his pictures to be of the highest value, not caring to paint except for great personages and those who were able to reward him properly." In fact, a list of Titian's portraits reads like a 16th Century *Who's Who.*

Oddly enough, the painting so largely responsible for catapulting Titian into this enviable position was a copy of another man's work. Although Titian had seen Charles and had presumably done some rough sketches of him, the details of the finished portrait were taken directly from a portrait of Charles done by the Austrian painter Jakob Seisenegger. It was not the first time, nor was it the last, that Titian used another man's work for inspiration. Once he did a portrait of Francis I of France from a medallion designed by Benvenuto Cellini. On another occasion he painted the aging Isabella d'Este as a young girl in her twenties *(page 94)* by using Francesco Francia's youthful portrait of the marchioness as a model. In a third instance Titian copied a lady's portrait to further a gentleman's political ambitions.

The lady was Cornelia, a young woman attached to the retinue of the Countess Pepoli, a Bolognese aristocrat. In 1529, when Charles V came to Bologna for his coronation, the Imperial party included an ambitious and handsome young man recently promoted to the position of chief political adviser to the Emperor, Francesco Covos. Francesco fell in love with Cornelia, and his infatuation was observed by Federigo Gonzaga, Marquis of Mantua. The equally ambitious Federigo—soon, in fact, to be made a duke—planned to consolidate his friendship with Covos by presenting him with two portraits of the Lady Cornelia. One was to be painted by Titian, and the other was to be sculpted in marble by a local artist, Giovanni Bologna.

Charles V was not a handsome man, but he felt that most portraitists made him look uglier than he was. In Titian the Emperor found an artist who finally did him justice by subtly minimizing his less attractive features and by emphasizing his characteristic strength, nobility and sincerity.

Portrait of the Emperor Charles V with His Dog, 1532-1533

Both artists were sent to the Pepoli palace to study their subject, but neither had much success. Giovanni Bologna fell sick and wrote the marquis to say that he was unable to do the job; he had, he said, a swollen cheek and a fever, and although he hated to disappoint his patron, it was impossible to act against the will of heaven. Within a few days Cornelia too became unwell and was shipped off to the country to recuperate. Titian wrote the marquis that he doubted he could have done anything good of Cornelia because he wasn't too well himself and, besides, was "suffering from the great heat." But he went on to assure the marquis that he had seen Cornelia, had been much impressed by her beauty and felt sure that he could paint a lifelike portrait of her—if he could be sent an existing portrait to copy. The marquis agreed to this proposition and sent Titian 100 *scudi* on account, a sum that inexplicably dwindled to 78.5 *scudi* on its way through the hands of the marquis's agents. He must also have sent along the requested portrait, for in September 1530 Cornelia's copied portrait reached her admirer.

Most people today would think it odd for a painter to copy another man's work, and odder still for a patron to receive it enthusiastically. To a modern patron of art, a painting is the artist's personal vision of what he sees; indeed, a modern portrait is as much a reflection of the artist's personality as it is a picture of the sitter—Picasso's portraits are in a sense paintings of Picasso. But the Renaissance, although it prized the individual artist's style, did not necessarily put much value upon the originality of his ideas. To an age that based its art on the artistic conventions of another age, on the classical art of Greece and Rome, it could scarcely have seemed strange for a man to base a painting on the work of another man.

Artists themselves thought nothing of copying for whatever reason and from whatever source suited their purpose, and one of the sources they copied from most frequently was, of course, classical art. Having accepted classical ideals of beauty and form, it seemed only natural to turn back to Greece and Rome for their models. Italy, with its wealth of ancient sculpture, abounded in examples. Another source, equally rich, was a collection of hundreds of engravings, illustrating stories from Greek and Roman mythology. It was produced by one of the finest engravers of the day, Marcantonio Raimondi, a man who had begun his career by making beautiful forgeries of the work of Albrecht Dürer. Raimondi's pictures were used by artists as a sort of handy reference file.

Sometimes these borrowed poses and compositions were used not only to evoke memories of the antique, but were also juggled into Christian religious settings. Many an *Entombment of Christ,* for example, can be traced back to the carvings of burial processions on Roman tombs, while many an Apollo was transformed into a David and many a Perseus slaying Medusa became David slaying Goliath. The anguished figure of Laocoön probably provided Michelangelo with the inspiration for his statue, the *Dying Slave,* which Titian may in turn have seen and used as his model for the figure of St. Sebastian in the altarpiece for the Church of St. Nazarius and St. Celsus in Brescia.

In addition to copying freely from the past and from each other, artists also copied themselves. When a painter found a particularly effective pose or design, he did not hesitate to use it over and over again, working and reworking the same thematic material. Botticelli repeatedly used the trio of female figures that are usually called the Three Graces; and the figure of a reclining nude, perhaps invented by Giorgione, was employed by Titian so many times that it has practically become his trademark. One artist conceived such a passion for a particular pose that it led to tragedy: Sansovino, it was said, required a certain young man to stand in a certain position so often that at length the wretched fellow fell into the pose compulsively at all hours of the day, holding it for longer and longer periods of time until at last he ceased to move at all, and so died.

Originality, therefore, was far less important to Renaissance patrons than accuracy and fidelity to nature. Men wanted their portraits to look exactly like them in every detail, and to look alive. Well-to-do merchants and wealthy aristocrats, men whose lives were spent in the pursuit of profits and power, wanted their portraits to reflect their material well-being and their place in society. There was more to this wish than self-pride; a man's portrait was closely bound up with the continuity of his family. He wanted his descendants to see him just as he was at the height of his fortune, surrounded by the symbols of his success—a man to admire and to emulate.

This kind of true-to-life likeness was called, aptly enough, a *contraffazione* or counterfeit, in the sense that a counterfeit dollar bill is a meticulous copy of a real one—and it was a term of approval. "What find I here? Fair Portia's counterfeit!" exclaims Bassanio in *The Merchant of Venice*. "What demigod hath come so near creation?" Speaking in the same vein, if less poetically, Hans Fugger, head of the great German banking house, announced that he wished to hire a portraitist "no matter whether he be otherwise skilled in painting, provided only that he be a master of counterfeiting." Among the multitudes of counterfeit portraits painted during the Renaissance few are more delightful than the wart-nosed, affectionate grandfather in Ghirlandaio's *Portrait of a Man with his Grandson* and few are more formidable than Antonello da Messina's scarred and craggy-jawed portrait of a man nicknamed *Il Condottiere.*

Jakob Seisenegger's portrait of Charles V—the portrait that Titian copied—was a typical counterfeit portrait, but Titian's copy was not. Seisenegger had produced a dry and precise record of the Emperor's appearance, whereas Titian's portrait, with its luminous colors and vibrant lights and shadows, was the portrait of a living man. It may not look like the real Charles, but it is a real person—the man in Titian's picture has vigor and dignity and a suggestion of the complexity of human character. It is, in fact, an idealized Charles and a picture of the Ideal Ruler—and this is the key to Titian's renown as a painter of people. Titian himself put it this way: "The painter ought, in his works, to seek out the peculiar properties of things, forming the idea of his subjects so as to represent their distinct qualities and the affections of the mind, which wonderfully please the spectator."

While he was working on the portrait of Charles, Titian also painted the portrait of another man of influence *(page 117);* Cardinal Ippolito de' Medici had, in fact, helped arrange Titian's introduction to the Emperor.

If this portrait of Charles V by the Austrian painter Jakob Seisenegger looks familiar it is because it is almost identical with the one by Titian shown on page 120. The Emperor wanted his portrait painted by Titian but could not arrange to sit long enough for the artist. But Seisenegger, whom Charles had "borrowed" from his brother Ferdinand, was able to complete this picture, which Charles gave to Titian to copy. Where Seisenegger's portrait is a meticulous inventory of the Emperor's features and costume, Titian's is a harmonious, living image of the man. It was the first of Titian's many portraits of the Emperor.

Ippolito's portrait is hardly what one would expect of a cardinal; he is shown in the dashing red velvet uniform of the Hungarian hussars. But then Ippolito was scarcely a model ecclesiastic. Pope Clement VII had made his bastard nephew a cardinal at the age of 17 to prevent the quarrelsome Ippolito from attacking his cousin Alessandro de' Medici, Duke of Florence. Apparently the Pope hoped that the clerical robes of the consistory would be a restraining influence on Ippolito's notoriously rowdy ways. It was a vain hope. Ippolito, who had always wanted to be a soldier, continued to prefer military dress. On one occasion he offended no less a person than the Emperor himself by appearing as papal legate at the Imperial court dressed in armor, an affront for which Charles had the cardinal arrested.

In 1532, when the infidel Turks invaded Hungary, Ippolito's preference for military pursuits earned him a papal appointment as commander of the defending forces. In the field Ippolito led his men quite well, but between battles his troops spent as much time plundering Hungarians as they did rescuing them. The cardinal was accused of laxness and even of taking a hand in the looting; for a brief time Charles V had him imprisoned. Eventually he settled down in Bologna, where he surrounded himself with a court so ostentatious that the Pope urged his nephew to dismiss some of his 300 courtiers. To this Ippolito replied grandly, "I don't keep them because they are necessary to me but because I am necessary to them."

The cardinal apparently became one of Titian's regular patrons. In a letter to Ippolito in 1534 the painter apologizes for his delay in sending the picture of a lady, explaining that the Cardinal of Lorraine had seen it and wanted a copy. Titian also sent Ippolito news of "my sons, Pomponio and Orazio," who were, he said, "well, and attending to their studies." Then, with more than a gentle hint, he added that "they are grown quite tall, and will, I trust, become great men by the favor of God and my patrons." Unfortunately for Titian's hopes for his sons, the cardinal's patronage lasted only five years. Less than a year after this letter, Ippolito was discovered in a plot to blow up his cousin Alessandro with gunpowder and seize Alessandro's dukedom. Forced to flee Bologna, the cardinal headed for Naples. But on his way south he fell sick and died of that common Renaissance malady, poison.

A soldier of a very different sort from the erratic Ippolito also sat for Titian. His name was Alfonso d'Avalos, Marquis of Vasto, commander of the Imperial forces in Lombardy. Sometime between 1532, when at the Emperor's behest d'Avalos went to war against the Turks, and 1535, when he helped to seize the port of Tunis from the Sultan's troops, Titian painted the marquis in a magnificent suit of armor. In 1539, after a grateful Charles made d'Avalos governor of Milan, Titian was commissioned to paint him again. This second portrait, which shows d'Avalos addressing his troops, commemorates an occasion when his powers of persuasion prevented a threatened mutiny. The date of the incident is unknown, but d'Avalos was clearly proud of it.

The pressure of other work and the apparently endless task of collecting outstanding debts kept Titian from completing this portrait until

August 1541. D'Avalos grew impatient and sometime during 1540 Titian placated him with a sketch. The finished work must have pleased him even more. The *Allocution,* as it is often called, shows the marquis standing before his soldiers in a proud pose reminiscent of those used to depict ancient Roman generals; the figure of Trajan, for example, appears in a similar stance on the column in his honor in Rome. Reinforcing this allusion to antiquity, the page boy next to d'Avalos is dressed in an outfit vaguely like the skirted battle dress of Roman warriors. D'Avalos himself wears a suit of contemporary armor, borrowed by Titian from a man in Brescia—in return for a promise to paint the Brescian's portrait.

In this painting is another portrait, an incidental one. Among the bearded soldiers listening to d'Avalos is the face of Titian's friend Aretino. "Whoever thinks me a flatterer," wrote the publisher Marcolini to Aretino, "let him look at your likeness in the picture where your more-than-brother Titian has painted in a most natural manner Alfonso d'Avalos haranguing the army like Julius Caesar in act and form. Milan runs . . . to look at you as a divine and most worthy image." It was not the first nor the last time Aretino served Titian as a model. Sometimes he was an anonymous face in a crowd, as in the *Allocution;* sometimes he was the subject of a formal portrait. In a large *Ecce Homo* painted in 1543 for a Flemish merchant living in Venice, Aretino was the model for Pontius Pilate; this painting is apparently rich in portraiture for it is also said to contain the faces of Charles V and Sultan Suleiman, as well as Titian's daughter Lavinia and Titian himself.

Aretino presented his three formal portraits by Titian to aristocratic friends, and two still survive. One shows the sharp-tongued author as a man well along in years, when he had stopped dyeing his beard and had resigned himself to old age. The other *(page 100)* shows him in the robes of a courtier. This painting went as a gift to Cosimo de' Medici, the new Duke of Florence. In a letter to the Duke, Aretino apologizes for certain parts of the portrait. Titian, he complains, has done a sketchy job on his costume. "If I had paid him more, truly the cloth would have been shining, smooth and stiff, like satin, velvet and brocade." In the portrait Aretino wears an elaborate gold chain, worth 600 *scudi,* given to him by the King of France. It weighed eight pounds and was hung with little gold serpents' tongues enameled in vermilion to represent poison. "His tongue uttereth great lies," reads the inscription—very possibly a subtle reproof by Francis of Aretino's excessive flattery.

Among the most important men Titian painted were the Doges of Venice, whose portraits he executed as part of his commitment to the Venetian state. But he seemed no more scrupulous about attending to this official task than about completing the Council Hall's long-delayed battle painting: many of these portraits were apparently sketched by Titian and finished by his assistants. One Doge, Andrea Gritti, fared better than others; his is one of Titian's most celebrated portraits. The Doge's stern visage peers out at the viewer like that of some old sea dog. Despite his formidable appearance, Gritti was a great friend of artists. It was he who encouraged Aretino and Sansovino to settle in Venice after their flight from Rome, and he was also most generous to Titian. He bestowed the post of Chancellor at Feltre upon Titian's brother-in-law, and in 1525 appointed Titian's elderly father inspector of mines for the region of Cadore. Several years later, after his father had died, Titian painted a portrait, which may be the old man's memorial; the grizzled *Old Warrior* wears the uniform of the Century of Pieve in which Gregorio Vecellio once served as captain.

One of Titian's most unusual portraits was a painting of a little girl, Clarissa Strozzi, the daughter of Roberto Strozzi, a wealthy Florentine whose anti-Medici sentiments had forced him to leave the city for a time. While in residence in Venice, Roberto commissioned Titian to paint Clarissa's portrait. The painting returned to Florence with the family sometime later and hung in the Strozzi palace there for about 300 years. In the middle of the 17th Century a little boy who was a guest in the palace saw the painting, fell in love with it and never forgot it. Years later, when he was in his sixties, Count Lorenzo Magalotti wrote to Clarissa's descendant, Leone Strozzi, with a touching request:

"I think that a famous original picture of Titian's must be there, at your house. It is a fine little girl standing, dressed in white, I think, with a watch hanging by a gold chain from her waist-ribbon. . . . Now hear me. I wish to have a copy of it." The count went on to explain that he first saw the portrait when he was about four years old, and "in the hands of my mother's usher with whom I was in the gallery of St. John the Baptist on the day of the festival, and it never left my mind so much did it please me at that age . . . every part of it is so fixed in my mind that I could paint it. . . . And add the following to the other proofs of the terrible ascendancy which this little girl had over my fantasy. You may suppose that when I was in the church of St. John no one thought of telling me it was by Titian. I remember well that the usher, seeing me look at it in ecstasy, said to me in a formal tone, 'That young lady is by the hand of the most skillful man that ever lived,' and I remember reflecting that there must be a great difference in painters."

It is nice to know that the count got his wish. Not all of the men who coveted Titian's portraits were as modest or courteous in their requests. Francesco Covos, Charles V's adviser, once asked the Duke of Ferrara to part with his own portrait by Titian; the Emperor, Covos suggested, might be better disposed toward Ferrara if he were to receive such a "gift." In vain the Duke's spokesmen protested that the painting no longer looked like him, that he was now much older; Covos was firm. Within two weeks the Duke's portrait was packed up and sent off to Charles, never to be returned. Understandably unhappy over the matter, the Duke asked Titian in 1534 to make a copy of this painting, and when he died later that year, his son Ercole urged Titian to finish it—sending him a 50-ducat down payment to seal the bargain. The copy must have been as fine as the original, for Ercole's agent wrote the new Duke from Venice that his father's portrait resembled the first portrait "as water resembles water."

Through his friendship with the ruling families of Ferrara and Mantua, Titian acquired the patronage of another Italian prince, Francesco Maria della Rovere, Duke of Urbino. The della Rovere were noted for their enlightened rule; one historian recorded that they "erected buildings, furthered the cultivation of the land, lived at home and gave employment to large numbers of people; their subjects loved them." Francesco was married to Eleonora Gonzaga, sister of the Duke of Mantua, and both husband and wife maintained a lively interest in the arts. But Francesco was also a man of action. Like many short men he was quick to take offense and equally quick with his sword; it was rumored that once he had even skewered a cardinal—for what breach of faith, no one knows.

In 1537 Francesco was appointed to lead an army representing Venice, the Vatican and the Holy Roman Empire against the Turks. Actually the army never materialized, for the three allies soon had a falling out. But sometime in 1538 Titian painted Francesco's portrait *(page 96)* in the armor he might have worn into battle on behalf of the coalition. The Duke holds the baton of command, and the standards of the three allies rest against the wall behind him. Judging from a sketch, Titian originally planned to make this a full-length portrait but changed his mind and painted it half-length—possibly to make it match the proportions of a half-length portrait of the Duke's wife painted around the same time.

In September 1538 the Duke fell inexplicably ill—as so often in that period, poison was suspected—and the following month he died. During his brief period of patronage, however, he purchased from Titian a number of paintings that were cherished by the della Rovere family for many years. Among them, according to Vasari, were a *St. Mary Magdalen* "with disheveled hair," and a group of portraits of Charles V, Francis I, three popes and Sultan Suleiman. Also in the della Rovere collection was a Titian painting purchased by Francesco's son and heir, Guidobaldo. It is one of the most delectable nudes ever put on canvas. The painting *(pages 98-99)* is usually called the *Venus of Urbino,* although Guidobaldo and his Venetian agent seem never to have referred to it as anything but La Donna Nuda, the Naked Lady.

The *Venus of Urbino* is not only a marvelous painting; it is also a splendid example of Titian's particular cast of mind. Superficially it is very like Giorgione's *Sleeping Venus,* produced some 30 years before, but in mood the two Venuses are worlds apart. Giorgione's goddess is pure spirit, an idyllic creature in a dream landscape; Titian's goddess is alluringly human, a woman lying on a bed. Giorgione's nude is truly nude, woman as nature made her; Titian's nude wears a little ring, a bracelet and earrings, and her servants in the background are selecting her clothes—all of which makes her lack of clothing seem much more voluptuous.

Titian was evidently much taken with the model who posed for the *Venus of Urbino,* for he painted her at least four other times. In one work she is the *Girl in a Fur,* in another *(page 97),* also painted for the della Rovere family, she is a woman in a blue and white gown, called simply *La Bella.* She appears a third time as the *Girl with a Feather in Her Hat,* and she is also probably one of the attendant women in a stupendous religious painting produced by Titian between 1534 and 1538, the *Presentation of the Virgin.* The identity of the lovely girl is unknown. Bernard Berenson thought she might be Eleonora Gonzaga as a young girl, and the Italian art historian Leandro Ozzola believed her to be Isabella d'Este. Other

This working sketch by Titian, made as a study for the formal portrait of the Duke of Urbino shown on page 96, conveys a feeling for the man as stumpy, powerful and balding—an effect somewhat different from the finished product. This rare drawing—one of only a few preparatory sketches that can be definitively connected with Titian's paintings—reveals that the artist may have originally considered a full-length portrait, or may indeed have painted one and then cut it down. The ruled squares on the sheet were a common device to aid an artist in transferring his design to the larger format of the canvas.

scholars have suggested that she was Titian's mistress—or Guidobaldo della Rovere's. Possibly all these guesses are wrong, and she was simply a girl with a good figure who passed briefly through Titian's life.

The *Presentation of the Virgin,* in which this young charmer appears simply as a bystander, is a huge canvas covering one whole wall of the refectory in the monastery of Santa Maria della Carità, a building that has since become Venice's famed art gallery, the Accademia. The painting illustrates one of the most popular religious themes in Renaissance art; Jacopo Bellini left behind a drawing of it, and Carpaccio painted it at least twice. The story of the Presentation comes from one of the apocryphal books of the Bible, the Gospel of the Birth of Mary, and tells of Mary's presentation to the temple at the age of three to serve as one of the temple virgins. According to the ancient account, the child astonished everyone by racing up the steps of the temple with such eagerness that it seemed she must have some foreknowledge of her sacred calling as the Mother of Christ.

In Titian's painting the watching crowd includes not only the face of La Bella—as his unknown model is often called—but reputedly the faces of several members of the Venetian Signoria as well as the face of the Grand Chancellor of Venice. Titian's setting is an elaborate architectural rendering of pillars and arches giving onto a distant view of sharp-peaked mountains like those of his native Cadore. The lower half of Titian's canvas is broken by two doorways leading into the hall where the painting hangs. One was apparently cut after the canvas was in place, but the other must have been there in Titian's time, for he ingeniously incorporated it into his picture by surrounding it with a painted masonry frame.

Three other monumental religious works, all begun by Titian in 1542, were done to the accompaniment of an angry lawsuit over money, one of the endless financial battles that studded Titian's career. Commissioned by the brothers of the Church of the Holy Spirit in Isola and installed on the church ceiling two years later, the themes of the three paintings were Cain's killing of Abel, Abraham's sacrifice of Isaac and David's victory over Goliath. Previous to this project, however, Titian had painted an altarpiece for the church, *Descent of the Holy Ghost,* which the brothers had subsequently rejected, claiming that it had darkened and discolored. Titian was unwilling to replace it until he had been paid, and the lawsuit followed. For more than a decade the matter dragged on, until finally, about 1555, the litigants came to terms and Titian painted a second altarpiece for the brothers, more to their liking.

Another argument over money was resolved more speedily. In 1537, at the request of the nuns of the Church of St. Mary of the Angels on the island of Murano, Titian had painted an altarpiece of the *Annunciation.* He had set a price of 500 *scudi* for the work, and when the nuns objected that the figure was too high, he sent the painting instead, at Aretino's suggestion, to the Empress Isabella, wife of Charles V. The Emperor was so delighted that he gave Titian 2,000 *scudi* for it. But this handsome windfall was almost canceled out by another financial difficulty. That same year the members of the Venetian Council finally lost patience with Titian for his failure to produce the long-delayed battle scene for the

Council Hall, and they issued a decree. Titian, it said, had held a broker's license and drawn a salary without performing as promised: "It is proper that this state of things should cease, and accordingly Titian is called upon to refund all that he has received for the time in which he has done no work." The amount involved was 1,800 *scudi,* a formidable sum even for a well-to-do painter.

The threat brought results. Within a short time Titian was hard at work on the battle scene, and by August 1538 it was completed. Unhappily the painting was lost, along with many others, when the Doge's Palace was gutted by fire in 1577, and no one has ever been quite sure what battle it depicted. Lodovico Dolce referred to it simply as "the battle," and Vasari thought it represented a rout of Venetian troops at Chiaradadda, a defeat of which there seems to be no record—and which seems besides an inappropriate choice of subject to decorate the Council Hall. Sansovino, who as a friend of Titian's ought to have known what the painting was about, said it portrayed the battle of Spoleto, another obscure military engagement that does not seem to have been noticed by history. Later scholars were sure it depicted the battle of Cadore, fought in 1508, a victory for Venice over the forces of the Holy Roman Empire. The evidence for this claim is an engraving of the painting by Giulio Fontana entitled *Titian's Battle of Cadore.*

In any case, the completion of the painting settled the troubles between Titian and the Venetian state. A year later, in August 1539, the Council restored his broker's license, and the painter once again enjoyed official esteem. Midway through his life—Titian was now in his fifties—he stood at the very top of his profession, secure, serene, far above the petty rivalries of other painters. In a self-portrait *(page 6)* painted a decade later, Titian appears as a man who is monarch of his domain. Rich materials cover his broad chest and ample paunch; the gold chain of knighthood hangs in a careless tangle around his neck; the gaze is proud. It is a portrait that radiates solidity, power and assurance—as much an idealized rendering of the painter as Titian's portrait of Charles V was an idealized ruler.

But there must have been another Titian, quite different from this man of substance. The other Titian was a man of simple tastes who liked a good joke with his friends, good food and wine, and a mild flirtation with a beautiful woman. "I wanted to conceal from you and Titian," Aretino wrote playfully to Sansovino, "that I supped with the beautiful Virginia, for I wished to keep your old age safe from voluptuous feelings"—and certainly it is clear from Titian's paintings of women that the advancing years had not dulled his sensuality. It is also clear that his attitude toward money was unchanged. Despite his comfortable surroundings he continued to petition his patrons for grants, pensions and annuities, meanwhile cannily investing his earnings in real estate. There is something in Titian's behavior of the rich man's mock-modest pretense of poverty, but it may go deeper than this. It may go back to the precarious life of Cadore. Like the peasant who can never be secure from natural catastrophe, Titian may have felt that he could never trust success. And so he went on ceaselessly coaxing life from his pigments, much as his forebears had coaxed life from the mountain earth.

Ruler of Titian's World

The overtowering political figure of Titian's day was Charles V, Holy Roman Emperor and ruler of Spain and the Netherlands. Not until Napoleon would there again be a sovereign known so widely in Europe and Africa; even the Americas heard his thunder in the relentless march of the conquistadors. In Europe, however, Charles was forced to act less as a conqueror than as a defender. In 1532 he led Imperial forces down the Danube against an army of Turkish invaders. Three years later he sailed across the Mediterranean to rout Arab pirates and reconquer Tunis. But his most implacable foe lay on the continent: during his 36-year reign, Charles fought France in many bloody and costly campaigns that dangerously depleted his treasury.

Surrounded by enemies and preoccupied with external conflicts, Charles neglected the hundreds of German political units that made up his Empire. It was there, however, that a most violent religious and political storm was gathering. From Martin Luther's daring critique of the Roman Catholic Church—the famed 95 Theses of 1517—a religious struggle mushroomed that split Christianity wide open and shook the foundations of monarchy throughout Europe. By the time Charles acted it was too late to stop the movement. In 1556, burdened by war debts and frustrated in his hopes for political peace in Europe and religious unanimity in the realm, Charles abdicated in favor of his brother Ferdinand.

Early on the morning of October 23, 1520, Charles V received the crown as Holy Roman Emperor from the Archbishop of Cologne in the Cathedral of Charlemagne at Aachen. In this stained-glass window from the Church of St. Gudule in Brussels, Charles and his wife Isabella kneel as his deceased grandfather, the Emperor Maximilian, passes on the symbols of rule—the orb and sword.

The stained-glass windows on this and the following pages were made from designs by Bernard van Orley, c.1530

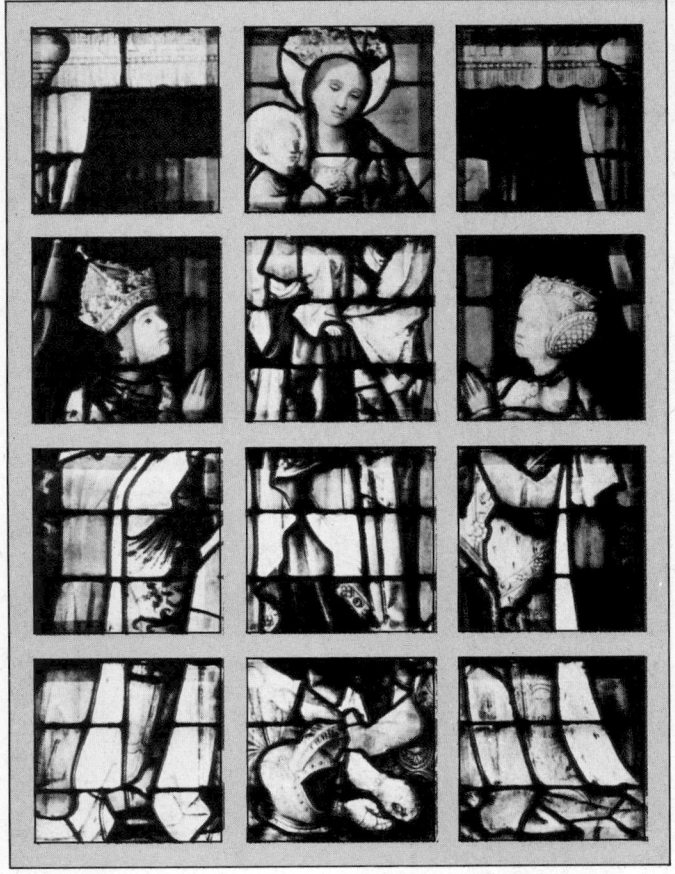

Maximilian I and Mary of Burgundy (grandparents)

"God has set you on the path towards a world monarchy," an adviser told young Charles. He had inherited vast lands and titles, which he would attempt to enlarge and consolidate by conquest in the Americas and by a series of diplomatically arranged marriages.

Philip the Handsome and Joanna the Mad of Castile (parents)

The stained-glass windows on these pages, from the Church of St. Gudule in Brussels, show the key individuals in his family. The genealogy spans a vital era of European history from before Charles's own birth in 1500 to the death of his son and heir Philip II in 1598. Of principal

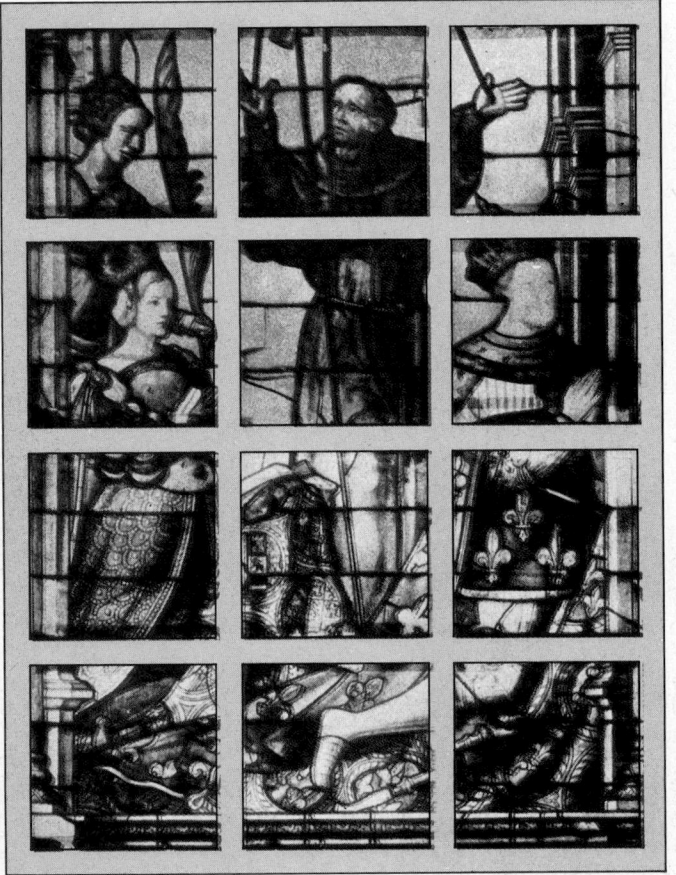

132

Francis I of France and Eleanor (sister)

John III of Portugal and Catherine (sister)

Philibert of Savoy and Margaret of Austria (aunt)

Louis II of Hungary and Mary (sister)

importance to Charles, in addition to his parents, were his grandfather, whose influence led to his election as Holy Roman Emperor, his aunt Margaret and sister Mary, who governed the Netherlands in his name, and his sister Catherine, whose marriage brought the Portuguese empire into the Habsburg sphere of influence. Charles was also related to great rival monarchies: France, through the marriage of his sister Eleanor; and England, first by the marriage of his aunt Catherine of Aragon to Henry VIII, and later by Philip's union with Mary, their daughter.

Ferdinand I (brother) and Anne of Hungary

Philip II (son) and Mary I of England

Although he had hoped that family alliances would thwart territorial ambitions within Europe, Charles spent a good part of his life at war. Much of the time he was in conflict with his brother-in-law, Francis I of France, who was fearful of encirclement. The threat was real, for in 1519 Charles had united the land surrounding France under Habsburg domination. Conflict between the two men became so intense and so personal that at one point Charles challenged Francis to hand-to-hand combat. Though the duel never took place, the gesture was fully in keeping with Charles's respect for courtly concepts of chivalry and manliness, traditions that had guided warfare for centuries, but which were becoming anachronistic.

With the growing use of guns and cannons from the 14th Century onward, the style of European warfare changed dramatically. Personal armor, for example, which had been of prime importance in old style close-quarter fighting, came to be used less for defense than for display of power and prestige. Thus, instead of declining, the armorer's craft rose to a fine art in the 16th Century, when exquisitely chased, engraved and gilded armor was worn in ceremonies, given as important gifts and offered as honors to gallant soldiers. Charles owned a vast collection of handsome armor, some of which is shown on these pages. Befitting its owner, much of it is made in a style imitative of Imperial Rome: dressed in the suit at the left—ornamented from the shoulders down to the calf protectors with fierce heads and savage satyr masks—Charles must have looked every inch a Caesar.

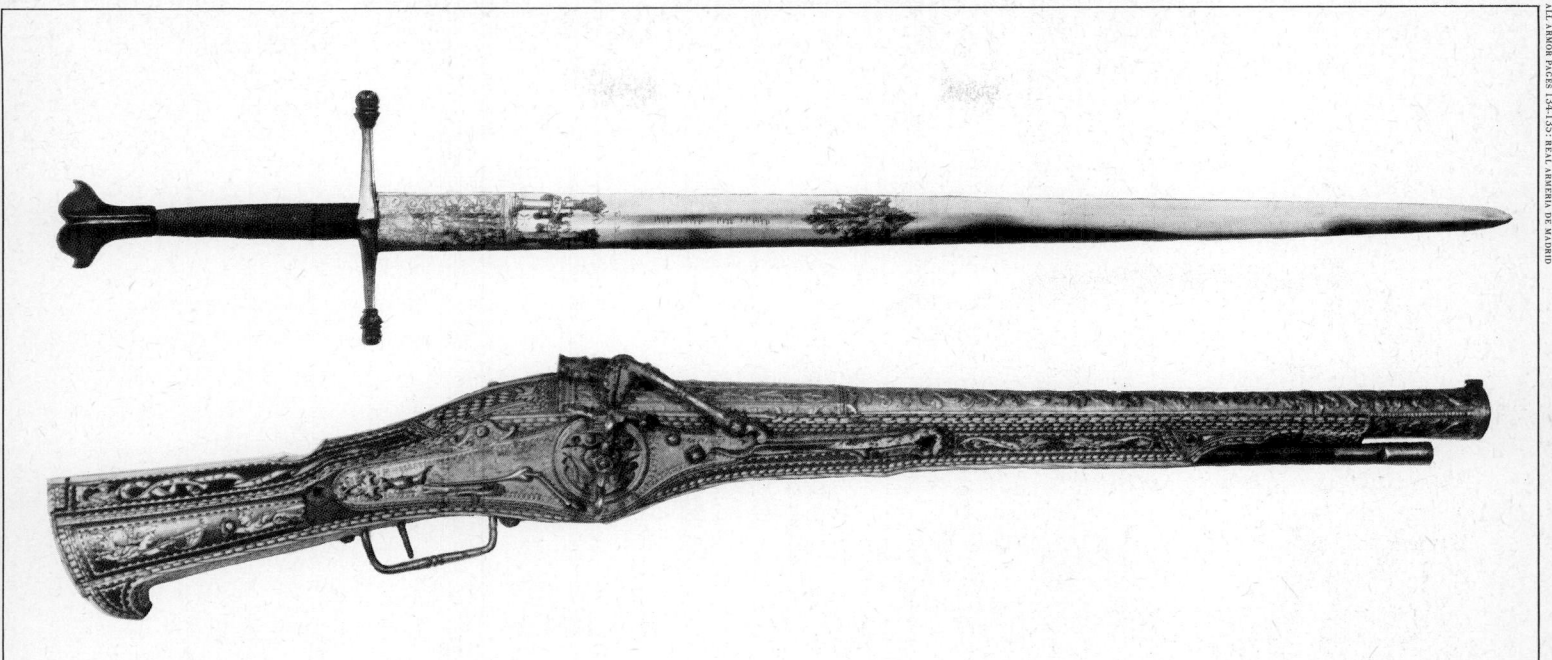

These weapons were in Charles's personal armory. The sword, with its yard-long blade, was used for hacking in man-to-man combat. The gun, a harquebus or wheel lock, was the first firearm that could be carried concealed and fired instantly. Swordsmen lamented the use of such weapons; the poet Ariosto called them "infernal tools." But Charles vowed to accept any weapon that would help him to serve God's cause.

This visorless gilded helmet mirrors the features of its regal owner. Displayed in relief on the neckpiece is the emblem of the Order of the Golden Fleece.

The chased lion head on this shield marks it as one that the Emperor used for display—combat shields were traditionally smooth-surfaced. Both shield and helmet were made by the renowned Milanese armorer, Giacomo Negroli.

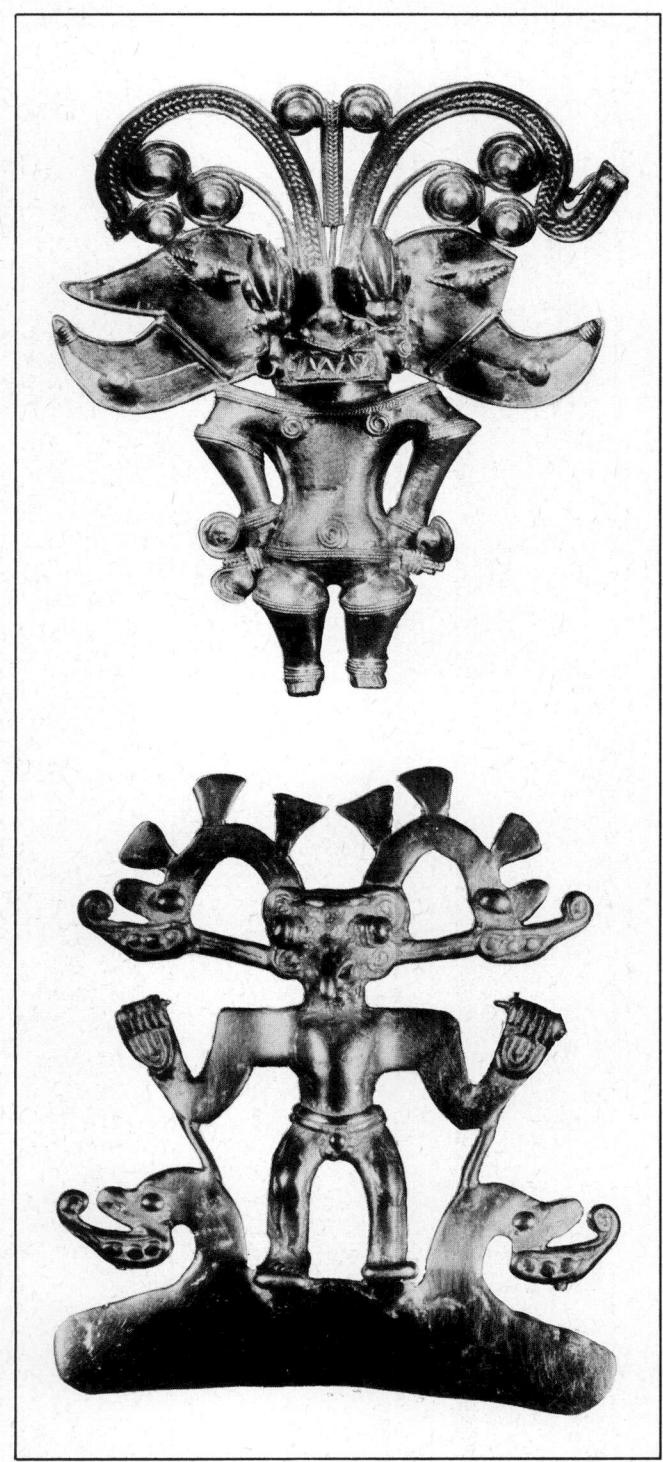

Charles required vast sums of money to prosecute his European wars; fortunately for him, his possessions in the New World would provide some of the wealth he needed. The expeditions of conquistadors greatly enlarged Charles's overseas domains, and the resulting subjugation of Mexico by Hernando Cortes and of Peru by Francisco Pizarro, brought significant treasure to the Empire. Although these

victories were brutal and violent, they were outstanding achievements of Charles's rule.

The adventurous Cortes, shown below presenting a petition to the Emperor, sent back to Spain tons of gold and silver, which he extracted by diplomacy and by force from the Aztecs of Mexico. Some of the treasures that were shipped to Europe from Peru were golden ceremonial

136

objects like the ones at the left below, beautifully wrought Incan ritual figures. The German artist Albrecht Dürer, on a visit to Brussels, inspected two rooms full of such objects that had just been sent from the New World. "All the days of my life," he wrote, "I have seen nothing that rejoiced my heart so much as these things, for I saw amongst them wonderful works of art, and I marveled at the subtle genius of men in foreign lands." Dürer's sensitive evaluation was not shared by all his contemporaries, who melted the gold and silver artifacts down and in some cases used the precious metals to make decorative ware for European royalty. The ornate gilded silver lid for a cup at the top, below, and the serving plate engraved in bands with Biblical scenes *(bottom)* are examples of such rich objects.

Illustration for the 1523 Zaragoza Edition of *Cortes's Second Letter to Charles V*

The greatest threat to Europe—and to Charles as the Protector of Christendom—came from the Muslim Turks and their Sultan, Suleiman. During the first decade of Charles's rule, Suleiman—shown at the right bearing a curved Turkish sword but dressed for war in the European style of full body armor—attacked the eastern borders of the Empire, occupying a chunk of Europe that extended all the way up the Danube past Belgrade.

In 1532 Charles arrived at the head of an army and drove the invaders away from Vienna. But the Turks then threatened in the south, where they occupied the North African city of Tunis, a strategic outpost dominating Mediterranean trade routes and perilously close to the Habsburg kingdoms of Spain, Sardinia and Sicily. In 1535 Charles—calling himself "God's standard bearer"—set out to recapture Tunis and invited Titian to join his retinue to record the battle. The artist prudently declined, but a Dutch engineer and artist, Jan Cornelius Vermeyen, went along. His sketches were later woven into a suite of tapestries. In the detail from one of them, below, the Turks are shown using primitive weapons— although they had long used gunpowder—to emphasize the superiority of the Christians. Charles routed the Turks from the city of Tunis in one of his finest personal triumphs.

SOLYMAN the most Magnificent, Emperour of the Turkes. He wan the Ile Rhodes and divers Ilands in the Mediterranean sea, overan Hungary, conquerd Babylon and the Countreyes of Mesopotamia: Tooke Strigonium, and won Alba Regalis, and at the seidge of Sigeth died, in Anno: 1567.

Albrecht Dürer: *Frederick the Wise, Elector of Saxony,* 1496

Lucas Cranach the Elder: *Portrait of Martin Luther,* 1533

Portrait of John Frederick, Elector of Saxony, c. 1548–1551

In addition to dangers from outside his Empire, Charles faced a menace within. So involved had he been with the French and the Turks that it was not until the late 1540s that he decisively turned to Germany. There, local princes were clamoring for an end to the political rule of the Habsburgs and religious control by the Vatican, both of which Charles, as Holy Roman Emperor, was pledged to maintain. As far back as the beginning of his reign, Charles had been threatened by princes who rallied to the support of Martin Luther in his rebellion against the Roman Church. Frederick the Wise, the Elector of Saxony, had given Luther sanctuary from papal and Imperial persecution. It was in 1547, against Frederick's nephew John Frederick, a member of the most powerful league of Protestant states and cities, that Charles was forced to act. At the Battle of Mühlberg, in April of that year, the Emperor crushed his opposition, captured John Frederick and took him to the Imperial court.

Titian, a guest at the court, painted John Frederick's portrait *(left)* for the Emperor. He also memorialized Charles's triumph in the painting at the right; wearing the symbol of the Order of the Golden Fleece around his neck and astride his Spanish steed, Charles is seen as a living St. George on his way to slay the Protestant dragon.

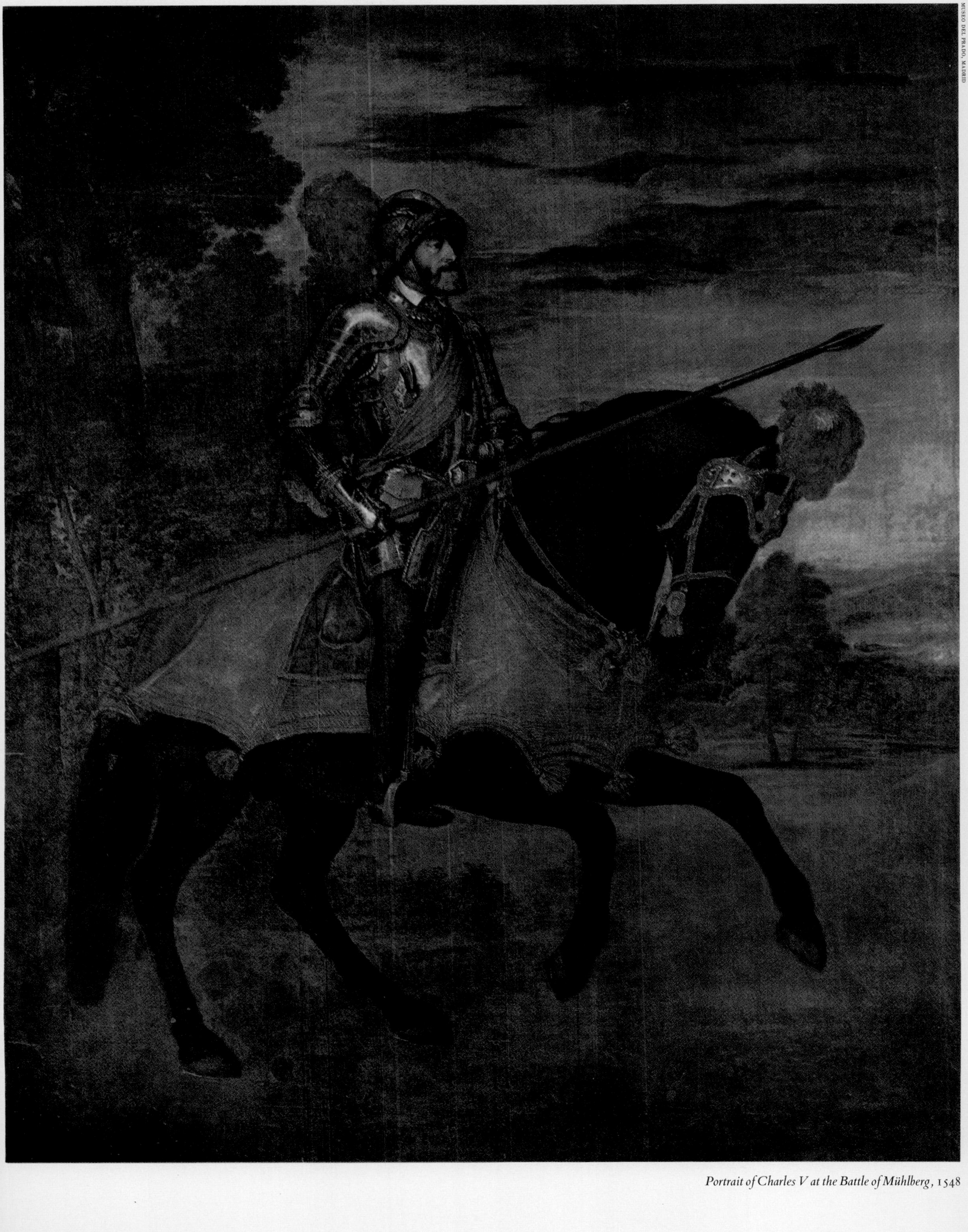

Portrait of Charles V at the Battle of Mühlberg, 1548

Portrait of Isabella of Portugal, c. 1548

Titian painted many portraits of the Emperor's family, among which are these elegant likenesses of his wife, Isabella, and his eldest son, Philip. Although Isabella had died before Titian made this portrait—he worked from an existing one—all her delicate beauty seems still in bloom. Charles's favorite pictures of his wife were by Titian; he kept them close by him throughout his life. Titian had more difficulty creating a handsome portrait of Philip; not only was the young man handicapped by the Habsburg nose, pouting lower lip and prognathous jaw, but also by a moody and sullen temperament. Yet Titian succeeded so well in lending the prince an air of aristocratic dignity that one of his portraits was alleged to have won the heart of Mary Tudor of England, Philip's intended bride.

Portrait of Philip II, 1551

51.

Danaë, 1553

Prince Philip became as devoted a patron of Titian as his father, commissioning portraits, religious pictures and a series of mythological subjects called *poesie.* Among these portrayals of classical legends—all of which concern the amorous adventures of gods, goddesses and the mortals with whom they sported —is this *Danaë,* one of the first of the *poesie* that Philip received.

Quite different from the repose and refinement of the *Venus of Urbino (pages 98-99),* which Titian had painted for Guidobaldo della Rovere, is the active eroticism of this scene. It was evidently very much to the taste of the young prince, whose sensuality, in the words of one historian, was "only equalled by his disregard for all that was good and kind in human nature." A Catholic zealot, a fanatical enemy of Protestantism and of all things that smacked of heresy, Philip was a private voluptuary who selfishly concealed the *Danaë* and other *poesie* that Titian created in a room designed for his personal pleasure. It is abundantly clear that the painting's classical theme—Zeus has disguised himself as a shower of gold to seduce the mortal *Danaë*—only thinly veils its unashamed erotic intentions.

Ecce Homo, c.1548

Mater Dolorosa, c.1544

At the age of 56, a weary Charles was relieved to hand over the Imperial crown to his brother Ferdinand, having recently resigned the Netherlands, the Spanish kingdoms, Sicily and part of Burgundy to his son Philip. At the bequeathal he turned to his peers and said, "Gentlemen, you must not be astonished if, old and feeble as I am in all my members, and also from the love I bear you, I shed some tears."

After bidding farewell to his son for the last time at Ghent, Charles sailed for Spain. In February 1557 he settled with his court at Yuste, west of Madrid, where he had built a villa next to a monastery famed for its sacred relics and noted as the burial place of many Spanish kings. Though far from fully retiring—he continued to mediate family squabbles—Charles lived quietly in Spain. He tried to find for himself the religious peace that he had failed to establish for his Empire. And as consolation and inspiration he surrounded himself with beautiful and spiritual art, among which were the moving *Ecce Homo,* the tearful *Mater Dolorosa* and the complex *Adoration of the Holy Trinity,* by Titian.

Titian's *Adoration* reveals a great deal about Charles's state of mind at the time, for the Emperor had instructed the artist very carefully about what he was to paint. The picture, which has been called a "manifesto of the Counter Reformation," is a final declaration of Charles's fidelity to Catholic orthodoxy. Principally, it represents the Trinity— Christ and God the Father are seated on high with the aura of the Holy Spirit about them. The existence of the Trinity had been denied by the Protestants, and the Emperor obviously intended the painting to refute this heresy and assert his own devout beliefs. He even had Titian include Noah (holding a dove and a model of the Ark), Moses (with the tablets) and other Old Testament prophets and patriarchs, because some theologians believed they had foreshadowed the notion of the Trinity.

Finally, Charles himself is shown kneeling prayerfully at right center with his crown beside him —his wife Isabella and son Philip pray behind him. Filled with his own deep faith and wearing a shroud in anticipation of death, Charles offers himself for final judgment. He seems to beseech God to forgive him his failure to stop Luther and squash the Protestant Reformation. It is a picture of humility, courage and dignity worthy of an Emperor.

The Adoration of the Holy Trinity, c.1551-1554

148

VII

The Emperor's Painter

In Titian's time three great powers dominated the Western world: France, the Church of Rome and the Holy Roman Empire. The Church, in addition, was also pre-eminent in the world of art. Few noblemen could match the rewards offered by the papal court. With myriad churches and palaces to be decorated, and a princely cluster of cardinals providing steady patronage, Rome promised artists wealth beyond measure. The city, moreover, was the very heart of the antique, a repository of gems from the past, filled with classical art and architecture. All roads led there, and all artists traveled them. Ultimately, Titian did so too.

As a young man Titian had been invited to Rome by Cardinal Bembo and had refused, preferring the pleasant life of Venice. In his middle age the invitation had been repeated, this time by Cardinal Ippolito de' Medici, but again Titian begged off, pleading the pressure of work. Perhaps he felt that Rome, which had already received and acclaimed Raphael and Michelangelo, was too crowded with talent, too competitive. Perhaps he sensed that the Medici were soon to lose the papacy. Indeed, the Medici Pope, Clement VII, died in 1534. His successor was a Farnese, Pope Paul III, and it was the Farnese who finally lured Titian to Rome.

Titian's first contact with the Farnese came in 1542, when 11-year-old Ranuccio Farnese passed through Venice on his way to school at Padua. Ranuccio was one of the five children of Pier Luigi Farnese, the son of Paul III. For all his tender years, the boy already held the post of prior of the Church of St. John of the Templars; at 14 he would be made a cardinal, just as his brother Alessandro and his cousin Guido Ascanio Sforza had been made cardinals by Paul III when they were in their teens. Nothing seemed more fitting, when the young prince stopped over in Venice, than that he should sit for his portrait by the city's leading artist (*page 117*). The idea seems to have been suggested by Cardinal Bembo, who had never ceased to admire Titian and had never given up hope of attracting him to Rome. Young Ranuccio's portrait was intended to bring Titian to the attention of the Farnese, and the stratagem worked. Before long Ranuccio's older brother Alessandro was offering Titian work in Rome and was promising him another benefice for his son Pomponio.

Titian's lifelong interest in the more erotic of the ancient myths and legends is fully evident in this brutal scene—the rape of the virtuous Roman matron Lucretia by the lustful Tarquin, son of a king. A late work, the picture shows Titian's deft control of mood in its violent composition—the flailing arms, the busy folds of drapery—and in the vibrant, freely applied color.

Tarquin and Lucretia, before 1571

Titian's concern for his three children was one of the driving forces of his life. When his daughter Lavinia was betrothed a number of years later, he moved heaven and earth to provide her with a handsome dowry of 1,400 ducats. When the younger of his two boys, Orazio, showed a talent for painting, he took him on as an apprentice and, before long, as a partner. As for Pomponio, the eldest of the three, Titian had destined him for a career in the Church, not as a humble priest but as the holder of a lucrative benefice. The position was one frequently sought by influential people for their sons, for it allowed the beneficiary of a parish to collect its revenues without fulfilling any of the parochial responsibilities. Out of the benefice, its holder withdrew a certain amount each year as a salary for the vicar who actually ministered to the parishioners' souls.

Pomponio had acquired his first benefice in 1531, when he was only six; the Duke of Mantua had granted him the living from the parish of Medole, and Titian had at once put his son into clerical garb—much as doting parents nowadays put small boys into sailor suits. Several years later Aretino wrote Pomponio a delightful letter in which he addressed him as "little Monsignor" and warned him that a priest must get accustomed to living on tithes. At 14 Pomponio received his second benefice, as canon of the Cathedral of Milan, through the good offices of his father's patron Alfonso d'Avalos, then governor of that city.

Now the Farnese were dangling still a third benefice for Pomponio before Titian's eyes, the income from the abbey of St. Peter at Colle near Venice. Titian was finding this difficult to resist. In September 1542 an emissary from the Farnese called on the artist and subsequently reported back to Cardinal Alessandro that he was sure Titian would "come and take service in the house of your Reverend and Illustrious Lordship . . . if you would acknowledge his talents and labors by the promotion of his son." The writer, a Roman scholar named Gianfrancesco Leoni, closed his letter by observing, "This man is to be had, if you wish to engage him. Titian, besides being clever, seemed to us all mild, tractable, and easy to deal with, which is worthy of note in a man as exceptional as he is." In the light of Titian's many lawsuits and his endless bickering over prices and payments, Leoni's appraisal seems somewhat implausible. But perhaps Titian, in the time-honored way of shrewd salesmen, simply put on an air of amiability the better to make his deal.

In 1543 Pope Paul left Rome and headed north on a diplomatic mission, accompanied by a group of church dignitaries that included Cardinal Alessandro. Paul hoped to arrange a meeting with Emperor Charles V, then returning from Spain to Germany by way of northern Italy. He wanted Charles to sell him the duchy of Milan for his grandson, Ottavio, the brother of Ranuccio and Alessandro. The plan fell through when Pope and Emperor could not agree on terms, but Paul's stay in northern Italy brought Titian into direct contact with the Farnese. When the papal party halted in Bologna, Cardinal Alessandro invited Titian to join him there as his guest. Titian, anxious to further the matter of Pomponio's benefice, accepted.

During this visit Titian was commissioned to paint two portraits of the Pope—one for the Holy Father himself and one for Cardinal Santafiore—

as well as a portrait of the Pope's son, Pier Luigi Farnese; he may also have produced at that time his portrait of Alessandro Farnese. Titian's first portrait of Paul is one of his most remarkable creations. It shows an old man, stooped, thin-lipped and gaunt, with blue-veined hands and a long, pendulous nose; everything about the portrait speaks of senility except the eyes, which are as crafty as the eyes of a fox. Pier Luigi's portrait reveals the face of a man equally cunning but lacking his father's strength. The younger Farnese left no vice untried, and Titian has painted him as a smooth, oily, treacherous voluptuary.

While in Bologna Titian discussed Pomponio's benefice with Alessandro Farnese and must have been disheartened to discover that the benefice was not really the cardinal's to give. It already belonged to Giulio Sertorio, archbishop of San Severino. Alessandro assured Titian that the archbishop would relinquish the position, but then he abruptly left Bologna without attending to the matter and without notifying Titian that he was going. In Venice, Titian wrote the cardinal that "the sudden departure of his Eminence had caused [Titian] to spend a bad night, which would have been followed by a bad day and a worse year if [the cardinal's secretary] had not come the next morning to say that Monsignor Giulio had ceded or promised to send the cessation of the benefice."

When the benefice was still not forthcoming, the Pope attempted to appease Titian by offering him the Office of the Papal Seals, a post long held by Titian's old comrade and co-worker Sebastiano del Piombo. The bright promise that Sebastiano had displayed as an apprentice to Giorgione in Venice had long since been dulled in the dissolute climate of Rome, and Sebastiano was now just another artist at the papal court. Titian was outraged at the suggestion that he deprive his old friend of his livelihood, and refused the Pope's offer. But he continued to press for the benefice for Pomponio. In March 1544, with Aretino's help, he composed a letter to Cardinal Alessandro's secretary, declaring that nothing would be more refreshing than "to hear that his Eminence had kept the vow made by the holy clemency of the Pope in respect of the benefice."

A month later Ranuccio Farnese wrote his brother on Titian's behalf: "I received a visit from Titian, who begged I would ask your Reverend Lordship to hasten the grant of the benefice to his son. Titian being a most estimable person, I beg to recommend him most earnestly." Then the youthful student, as if to reassure his older brother that he was not dawdling in Venice when he should be in school, added, "I leave tomorrow for Padua." In the same month Titian even wrote to the aged Michelangelo—whom he had once met briefly in Venice—asking him as a brother artist to drop a word for him in the right quarters in Rome. But nothing came of this.

At last, in 1545, Titian himself set out for Rome. What finally motivated him is uncertain. Perhaps it was the hope of nailing down Pomponio's benefice, perhaps it was the lure of the excellent commissions that might await him, perhaps he could no longer ignore the urgings of powerful friends, perhaps he simply wanted to see the Eternal City at long last. Whatever his reasons for going, he went in style. Guidobaldo, the new Duke of Urbino, paid for his journey and gave him an escort of seven riders. In fact, Guidobaldo made the quite magnanimous gesture of accompanying Titian as far as Pesaro. In Rome the painter was received like a prince. Pope Paul III gave him rooms in the Belvedere Palace, close to his own quarters, and Cardinal Alessandro assigned Giorgio Vasari to act as Titian's guide and companion. Vasari, then a brisk and ambitious artist of 34, knew everyone and knew all the gossip about everyone. Detractors called him a sycophant, but Vasari was later to turn his wide acquaintance and intimate knowledge into the series of biographies, the "Lives" of artists, that made him famous.

Titian threw himself into sightseeing. He told Cardinal Bembo of his joy at seeing such a feast of antiquity spread out before him, and wrote Aretino that he regretted not having come to Rome 20 years before. Aretino wrote back, warning him to be on guard against the promises of the Farnese and begging him not to stay away all winter. "I long for your return, that I may hear what you think of the antiques. . . . I want to know how far Buonarroti approaches or surpasses Raphael as a painter; and wish to talk to you of Bramante's Church of St. Peter. . . . Bear in mind the methods of each of the famous painters . . . contrast the figures of Jacopo Sansovino with those of men who pretend to rival him, and remember not to lose yourself in contemplation of the *Last Judgment* at the Sistine, lest you be kept all the winter from the company of Sansovino and myself."

Titian met often with his friend Sebastiano del Piombo, who had largely given up painting in favor of rich food and drink and merry company, and who apologized for himself by saying, "There are geniuses around now who do more in two months than I used to do in two years. Since these good people are doing so much, it is just as well that I am content to do nothing and thus leave them all the more to do." With Vasari, Titian spent long hours strolling through palaces and galleries, studying the works of classical artists as well as those of his contemporaries, especially those of his only two rivals for fame, Raphael and Michelangelo. Later he was to admit that his visit to Rome changed his own work profoundly. But Titian had not come to Rome simply as a tourist, and soon he was hard at work. He produced a second portrait of Cardinal Alessandro, another portrait of Paul III and a family portrait of the old pontiff *(page 116)* with his two grandsons—usually called his "nephews" out of deference to Paul's nominal celibacy.

In this extraordinarily revealing work Alessandro, in his scarlet cardinal's robes, stands to one side of the seated Pope, aloof and dignified; on the other side, Ottavio, in courtier's garb, bends obsequiously toward the old man; Paul, bent with age, peers up at this self-serving young man suspiciously—as if aware of Ottavio's plot to depose his own father as Duke of Parma and seize the duchy for himself. Some historians think that the unfinished quality of the painting— Paul's face and figure are done much more sketchily than those of his two grandsons—suggests that the Farnese stopped Titian in mid-painting, alarmed at its implicit disclosures. But Vasari reports that the Farnese were enormously satisfied with the picture, and indeed it seems unlikely that a family as powerful as these Roman princes would have cared in the least what a portrait revealed about their characters, their intrigues and their familial relationships.

Among other paintings Titian produced for the Farnese is a splendid *poesia* based on the legend of Zeus and the Greek princess Danaë, in which the god seduces the maiden while appearing to her in the guise of a shower of gold; the result of this dazzling liaison was Perseus, the slayer of Medusa. Titian's *Danaë,* like the *Venus of Urbino,* is a thoroughly alluring young woman. She lies on a couch against a red silk drapery, lightly veiled with a bit of white cloth; Cupid, at her feet, moves off to the right, looking back over his shoulder at the rain of golden coins falling from a cloud in the blue sky. But there are differences between Danaë and Venus that reflect the influence of Roman art and artists upon Titian. The figure of Danaë may have been inspired by one of Michelangelo's river god sculptures. Danaë, furthermore, is a creature of more substance than Venus; the shadows are deeper, the flesh more solid. Danaë rests upon her bed heavily; Venus floated upon hers. It would seem that Titian was affected by the weightiness and substance of the carved marble that he saw around him on every side.

Michelangelo, however, was not taken with this painting of Danaë, according to Vasari. When the painting was finished, he and Vasari stopped by Titian's studio in the Belvedere Palace to have a look at it. They praised the work to Titian's face, "as people do when the artist is standing by," but after they left Michelangelo remarked that, while Titian's manner and coloring were excellent, it was a shame that he had never learned how to draw, "for if this artist had been aided by art and knowledge of design, as he is by nature, he would have produced works which none could surpass. . . ."

The contrast between the Venetian and Florentine approaches to painting was never expressed more clearly than in this statement, a contrast that was in a way inherent in the nature of the two cities. Venice was lively, full of light, floating on water, ringing with the sounds of carnival. Florence was a city of fierce passions and intense civic pride; it burned its "vanities" in public to please a puritanical monk and found its noblest civic expression in the severely patterned façades of its palaces

Among the fewer than 20 authenticated drawings by Titian is this delicate landscape study. It seems likely that the sketch served as the model for the background of a woodcut depicting the sacrifice of Abraham, which was printed for Titian in about 1516. The artist must have made other such meticulous studies from nature for practice and as references for the landscape sections in his paintings, but if he did they have either been destroyed or remain to be identified.

and churches, the deeply incised shadows of its narrow alleys. Venice was all poetry, Florence all geometry. Venetian speech was easy-going and slurred, as exotic as St. Mark's Byzantine basilica; Florentine speech was pure and precise, as hard as the marble its sculptors loved.

In the work of Florentine painters, color was subservient to design; it was used to define carefully drawn compositions. In the work of Venetian painters, and especially in the work of Titian, design provides the rough framework upon which the painter explores color and canvas to their utmost limits. Brushing on his color layer after layer, building up structures of pigment, Titian produced brilliant depths, lights and darks that seem to glow from within the canvas; flesh that had the tone and illusion of life. He was also doing something else: he was organizing masses of color on canvas that were exciting to look at for themselves alone—an approach to painting that foreshadowed modern art and that neither Vasari nor Michelangelo, for all his genius, understood.

Despite Aretino's plea that he not stay in Rome all winter, Titian remained there until the following spring. In December Arentino wrote him of a terrible blow to their friend Sansovino. Among his many commissions for the Venetian state, Sansovino had taken on the design and construction of a new library opposite the Doge's Palace on the Piazzetta San Marco. The library, with its double row of arches and its wealth of decoration, has been called the crowning triumph of Venetian architecture. But on the night of December 18, 1545, shortly after the main hall had been vaulted over, the structure suddenly collapsed. Swiftly and mercilessly, without benefit of trial, the incensed Venetian government arrested Sansovino, threw him in prison, fined him 1,000 ducats and stripped him of his official post as chief architect of the Republic.

Sansovino's friends were in despair. "All the night instead of sleeping I walked about the room," wrote Aretino to Titian, "thinking what great disasters fortune had designed for so virtuous and honest a man, and condemning fate as extraordinarily capricious in making that very work which appeared so likely to be the tabernacle of the glory of our brother, become the sepulchre of his fame." Rallying to the sculptor's aid, his friends showered the Venetian authorities with letters, denunciations and countercharges. Finally they succeeded in having the calamity investigated, and Sansovino was exonerated. It was not the design that was at fault, but the severity of the winter weather, which undermined the library's structure. The Council restored Sansovino's position and rescinded the fine—although in characteristic Venetian fashion it held back 10 per cent of it, presumably to pay for the cost of processing his case.

In the spring of 1546 Titian went back to Venice, full of honors but still without the benefice for his son. Aretino had been right: the chief gifts of princes were promises. Yet Titian continued to address hopeful letters to Cardinal Alessandro Farnese in Pomponio's behalf, and in June 1547, when his old friend Sebastiano del Piombo died, he hinted that he himself would be willing to accept Sebastiano's vacated position in the Office of the Papal Seals. The cardinal seemed responsive to the idea, but before he had a chance to act Titian received a summons from another quarter. Charles V invited him to serve as official painter at the Imperial court in Augsburg. When news of this appointment got out, people flocked to buy Titian's work and be associated with him in any way. "It was most flattering," wrote Aretino, "to see . . . the crowds of people running to share in the productions of his art, and how they tried to buy pictures and everything that was in the house at any price." Titian penned a delicately humble note to Cardinal Alessandro, professing great sorrow at having to abandon his hope of working for the Farnese and explaining that the Emperor was really "forcing" him to come to Augsburg. In a master stroke of subtlety, he asked if Pomponio might be granted the Colle benefice to console Titian for not being able to serve the cardinal. The Farnese, anxious to remain on good terms with the Emperor, quickly handed over the benefice to the Emperor's favorite.

Packing up his painting materials and several pictures as gifts for Charles V, Titian set out north across the Alps at the beginning of 1548. He took with him his son Orazio and a young relative, Cesare Vecellio, who showed a talent for painting. Since Titian's last meeting with Charles the world had seen profound changes. The rumble of religious discord, signaled 30 years before by Martin Luther when he nailed his 95 Theses to the door of the castle church at Wittenberg, had now divided Germany and much of Europe into two bitter camps. Every political action was now somehow affected by the struggle between Protestantism and the established Church, and in Charles this struggle took on personal overtones. A devout Catholic, he was forced to seek some sort of accord with his Protestant princes in order to retain their allegiance in his struggles with his two greatest foes: France in the west and the Ottoman Turks in the east. The Pope, who disapproved of appeasing Protestants, criticized Charles in words that sometimes made Rome sound as if it were on the side of the Moslem Turks. Nevertheless, Charles persisted.

In 1544, when the German princes convened at the Diet of Speyer, Charles declared that Protestants and Catholics must be united in brotherly love—and thus gained sufficient support to launch a major attack upon France. In three months the Imperial armies had driven to within 25 miles of Paris, and Francis I was forced to sue for peace, ending the last of a succession of wars between Charles and Francis that had stretched over 23 years; within three years the French king would be dead. With France subdued, Charles quickly negotiated a treaty of peace with the Turks and was finally free to turn his attention to the religious problem within his Empire. Although he was still hopeful of mending the breach, his diplomacy had little effect. The Lutheran princes mistrusted him, and the Pope, behind the scenes, fanned that mistrust, seeking to drive a wedge between Protestants and Catholics. In the end, the Emperor found himself the ally of the Pope in a war against the dissident princes within his own realm.

The Vatican contributed a large sum of money to this alliance and an army under the command of Paul's grandson, Ottavio Farnese. Charles himself led the Imperial army, even though his doctor advised him against it; he was troubled by asthma and so crippled by gout that one of his feet would not fit into the stirrup. There was time for only a few bloodless skirmishes before winter set in, during which Charles suddenly found himself abandoned by his Roman ally. Paul had been playing for private stakes; he wanted the Emperor to cede the duchy of Milan to his son, Pier Luigi Farnese, and when Charles gave it instead to Ferrante Gonzaga of Mantua, the Pope ordered Ottavio to leave the field. Despite this setback, Charles renewed his offensive when spring came. In April 1547 he engaged his strongest Protestant foe, Prince John Frederick of Saxony, in a major battle at Mühlberg on the Elbe River. Charles scored a resounding victory and afterwards rode out proudly, lance in hand and clad in splendid gold-inlaid armor, to receive John Frederick's surrender. Titian painted him this way when he first arrived in Augsburg.

To this portrait of Charles on horseback *(page 141)*, armed, triumphant and vigorous, Titian later added a companion portrait, showing the Emperor in a very different guise. He is seated in an armchair in somber clothes, relaxed and weary, a private man rather than a public personage. Carlo Ridolfi later wrote that this portrait was hung facing a door and was so natural that those who passed by and glanced in thought they were seeing the Emperor in person. Charles also commanded Titian to paint a portrait of his noble captive, John Frederick of Saxony, a man so enormously fat that only one horse, a huge Frisian stallion, could carry him without buckling. Charles had brought John Frederick back to Augsburg with him and had placed him under house arrest in a mansion directly across the way from the Imperial palace. Titian painted him there, an immense, choleric man with bloodshot eyes and a cheek still marked by the gash he received in the Mühlberg battle *(page 140)*.

While at Augsburg Titian also painted a second portrait of Charles's beloved late wife, the Empress Isabella *(page 142)*. Like the first portrait he painted of her in 1544, it was based on several existing portraits of Isabella —for the Empress had been dead since 1539. One of these portraits was by the same Jakob Seisenegger whose work had provided Titian with a model for his portrait of Charles in 1532. When Titian delivered his first portrait of Isabella to Charles, Aretino wrote the Emperor that the picture ought to console him for Isabella's death, since "though God possessed one Isabella, Charles had the other." Charles must have been equally consoled by this second portrait of his Empress, for he kept it always near him for the rest of his life. It shows Isabella as a frail and delicate woman with fine features, red-gold hair and a marblelike skin, set off by a glowing red gown. She seems more alive in Titian's portrait than in the paintings that served him as models—paintings presumably done from life.

During his stay in Augsburg Titian also produced a whole bevy of portraits, since lost, of the Imperial family. He painted the Emperor's sister Mary, Queen of Hungary, and her niece, the Duchess of Lorraine; the Emperor's brother Ferdinand as well as each of Ferdinand's five daughters and two sons. The more he painted, the more Charles's admiration grew: the Emperor was once said to have remarked that he valued the acquisition of a new painting by Titian as much as a new province. Probably the remark was apocryphal; Charles was not the sort of man to equate art with territorial expansion. Still, he obviously cherished Titian, and his favors and compliments eventually drew jealous words from envious courtiers. On one such occasion, according to Ridolfi, the Emperor silenced the critic by observing pleasantly but firmly that he had many courtiers, but only one Titian.

Charles also showed his gratitude in more practical ways. While the painter was in Augsburg, the Emperor granted him a second pension drawn on the treasury of Milan, doubling the amount of one previously granted him in 1541. This pension, however, like others awarded to Titian, never actually got paid. Despite the Emperor's insistent reminders, Milan was reluctant to part with the money. In September 1549 Titian sent the Milanese governor a portrait of Charles and a graceful note, thanking him "for the courteous and friendly way in which your Excellency proffered . . . to obtain the payment of my pension." But by February 1551 his tone had changed. In a letter to the governor's secretary, Titian asked that gentleman to use his influence to "squeeze some money out of the grasp of the treasury," and referred to the promised grant as "my passion instead of my pension."

Titian returned home to Venice briefly in October 1548, only to be called back to Augsburg again in November 1550. He went willingly enough, since it was the Emperor who summoned him, but he wrote to his friends complaining of "this frigid zone where we are all dying of the cold." On this visit to Augsburg, Titian took with him letters from Aretino to various people at the Imperial court, asking their help in a curious endeavor. Aretino, who was so often called the "scourge of princes," himself wanted to become a prince. Now growing old, he longed for the peace and security of the Church: he wanted to be named a cardinal. Perhaps Aretino deemed himself just as worthy of wearing the red hat as those he had once mocked for their intrigues. Was he not known as the "banker of the poor" for his habit of dispensing coins to old people and children as they ran by on the canal banks alongside his gondola? Titian wrote him from Augsburg that many important men, including the Emperor, were anxious to do their best for him, but in the end nothing came of it—and perhaps it was just as well. No cardinal's hat was ever widebrimmed enough to hide the sins of that old reprobate, whose home at one point housed no fewer than 12 mistresses.

Among Titian's commissions on the second trip to Augsburg were several portraits of the Emperor's son, Prince Philip of Spain, who was soon to marry Mary Tudor, Queen of England. One of these portraits of Philip by Titian was, in fact, instrumental in convincing Mary to accept the suit of the Spanish prince. When the marriage was proposed to the solidly British Mary, she expressed some misgivings about marrying a man 11 years her junior who might be "inclined to voluptuousness." To reassure her, the Habsburgs sent her a portrait of Philip—probably the first portrait of the prince Titian did when he came to Augsburg—and Mary was reportedly "deeply enamoured." The painting shows Philip in armor at age 23, slender and a little awkward, for Philip was no athlete (page 143). The face is handsome enough, but there is a hint of the brooding melancholy that in later years caused him to rule Spain from the seclusion of the palace monastery of the Escorial, supervising every detail of the administration of his realm as scrupulously as a certified public accountant.

Although Philip was a libertine, he was Spanish enough to hide it behind grave and courtly manners. In years to come he would be one of Titian's most important clients, commissioning from him splendid religious paintings for the royal palace in Madrid, paintings suitable to a monarch who considered himself a defender of the Faith. At the same time, for Philip's own private collection, Titian would be painting some of his most majestic *poesie,* replete with nude gods and goddesses.

Titian's early portrait of Philip was one of several done to the order of Charles V as part of the Emperor's scheme to improve Philip's public image. Charles was desperately weary, old before his time. The cares of his realm, its debts and religious disturbances, sat heavily upon him. He had ruled for 35 of his 50 years, and he was thinking seriously of abdicating. His logical successor was his brother Ferdinand, ruler of Austria, but Charles was thinking ahead. Philip would be a logical candidate to succeed Ferdinand when Ferdinand relinquished the Imperial crown, and Charles wished to groom his son for this contingency. As a future candidate, Philip already had one thing in his favor—the enormous wealth of the Spanish Empire. But he knew nothing about the German people he might one day rule, and it was this flaw that Charles hoped to remedy when he brought Philip to Augsburg. The plan was not entirely successful. Although Philip learned to ride German horses, dance German dances and drink German beer, the taciturn Spanish prince never showed much talent for *Gemütlichkeit.*

Sometime after May 1551 Titian bid Charles a final farewell—he was never to see the Emperor again—and by August was back in Venice. From this time onward an air of sadness seemed to pervade the painter's life, although he still had 26 years left to live. His sister Orsa, who had been his housekeeper for 20 years, had died in 1550. Five years later his daughter Lavinia married and left his house. His son Orazio was busy with his own paintings, and his other son was a sore disappointment to his father. Pomponio showed a taste for the lowest sort of high living; in 1547 Titian had written of him bitterly that "as a boy he deprived wives of their husbands and now that he is a man he takes sons from their fathers." In 1554 Aretino took Pomponio to task for his spendthrift ways, and shortly thereafter Titian asked the Duke of Mantua for permission to transfer Pomponio's benefice at Medole to a nephew he thought more worthy.

Year after year, Titian tried to collect the various grants promised him by his patrons. In 1554 he wrote Charles V to report that the 200-*scudi* pension on the Milanese treasury still had not been honored, nor had an Imperial order permitting him to draw upon the grain supplies of Naples. He also pointed out that a 500-*scudi* Spanish pension awarded him by Philip was proving to be equally unproductive. Considering his poor health and the expense of his daughter's marriage, Titian hoped that "the greatest Christian Emperor that ever lived would not suffer his orders to be ignored by his ministers." Along with this letter Titian sent Charles a painting of the Virgin, probably the *Mater Dolorosa (page 146),* whose grieving countenance, he said, "reflects the quality of my troubles."

But these cares must have seemed as nothing compared with the great blow that fell in 1556. On the evening of October 21 his good friend Aretino, entertaining at a late dinner in his home, tipped back his chair to laugh at a joke, lost his breath and began to cough, and within an hour was dead. A final bitter jest is attributed to him, probably untrue but certainly characteristic. As they oiled his body for the last rites, Aretino is supposed to have regained consciousness long enough to mutter, "Now that I'm greased, guard me from the rats." No record survives of how Titian marked his passing, but surely the painter's world must have been darkened by the death of his old companion.

In the following February another event took place that must have added to Titian's sadness. In a strange and almost unprecedented move, the Emperor Charles V retired to a monastery. Dividing the Habsburg lands between his brother Ferdinand, who was soon to be the new Emperor, and his son Philip, whom Charles had made King of Spain and the Low Countries, the ailing Emperor entered the monastery of San Gerónimo de Yuste in the mountains of central Spain. There, surrounded by a retinue of some 50 attendants, he lived out the remaining year and a half of his life in a style that was simple if not precisely austere. When he was not at his devotions, Charles divided his time between his favorite sport, hunting, and his favorite food and drink—smoked eels, anchovies, sausages and iced beer—a diet that did his gout no good whatever.

Charles had brought with him from Augsburg a few of his most cherished paintings, among them Titian's portrait of his Empress. But Charles' eyes turned most often to a painting called *The Adoration of the Holy Trinity (page 147)*, a spectacular religious scene that Titian had sent him in 1554. Its focal point is the Trinity, suspended above the earth in a luminous cloud that becomes, at the bottom of the canvas, a ring of adoring faces of saints, angels, patriarchs and prophets. The Virgin approaches the Trinity on one side to intercede for the Imperial family, who kneel on the other side; Charles, Isabella and their children are dressed, somewhat bizarrely, in their burial clothes. Seated in front of this painting, the lonely, gouty old monarch contemplated eternity with tears in his eyes and waited for the end.

On September 21, 1558, Charles died. Titian, now at least 70, continued to paint. Although his workshop, staffed by five or six painters, took over most of the burden of making copies and filling routine details, Titian himself was still to produce many marvelous canvases. The world was changing, and tastes were changing with it, but he was still the acknowledged master. "In Titian alone," wrote Lodovico Dolce, "all those excellences are collected together in perfection which are found dispersed in many others. . . . Titian has not shown any vain desire of beauty but a propriety of coloring; no affected ornaments but the modesty of a master; no crudeness but the tender fleshiness of nature." Such tributes usually signal the close of a man's productive life, but for Titian the creative process was far from ended. Although two brilliant young painters—Tintoretto and Veronese—were rising in Venice to challenge his supremacy, the late canvases of Titian still surpassed theirs in depth of feeling. In the two decades still left to him Titian was, in fact, to produce a kind of painting advanced beyond the understanding of his own time.

This catafalque decked with hundreds of candles blazed in memory of Charles V in a Brussels church in December 1558, two months after his death. Although Charles was buried in Spain, where he had died, his son Philip ordered elaborate funeral services in Brussels as well because Charles had reigned over the Low Countries. The catafalque stood in the Church of St. Gudule, where imposing stained-glass windows *(pages 131-133)* of Charles and his entire family loomed over the mourners.

Titian's Successors

By the mid-16th Century, Titian had enjoyed decades of fame. Now he was joined at the forefront of Venetian painting by two much younger men, Tintoretto and Veronese. Both these artists had grown up in an atmosphere dominated by this titanic master, but if they expected that the aging Titian would retire gracefully, leaving the field to them, they were mistaken. The old man continued to create for another quarter century, while each of the others was displaying his own vastly rich talents.

Titian, in the last 10 or 12 years of his life and secure in his renown, painted as much to please himself as to satisfy patrons. Still a vigorous personality even in his eighties, he expressed passionate feelings in his works—in an age that still tended to regard a painting as a piece of fine craftsmanship rather than an outlet for personal emotions. Titian covered his late canvases with quick, broad brushstrokes, using shimmering colors to create blurred, evocative shapes rather than carefully delineated forms.

Tintoretto, an innovator himself, moved toward, and helped to establish, a new way of painting, which combined love of rich color and a vigorous sense of drama.

Unlike Titian and Tintoretto, Veronese was content to evoke the elegance and pageantry of contemporary life. His work, skillfully executed and abounding with bright light and dazzling color, brought 16th Century Venetian painting to a resplendent climax.

Titian's highly emotional late religious pictures are often dimly lit and softly colored. The somber scene at the right looks as though it might have been painted with colored smoke, a pictorial effect that gives the hazy, indistinct appearance of the figures of Christ and his tormentors a vital power.

Christ Crowned with Thorns, c. 1570

SCUOLA DI SAN ROCCO, VENICE

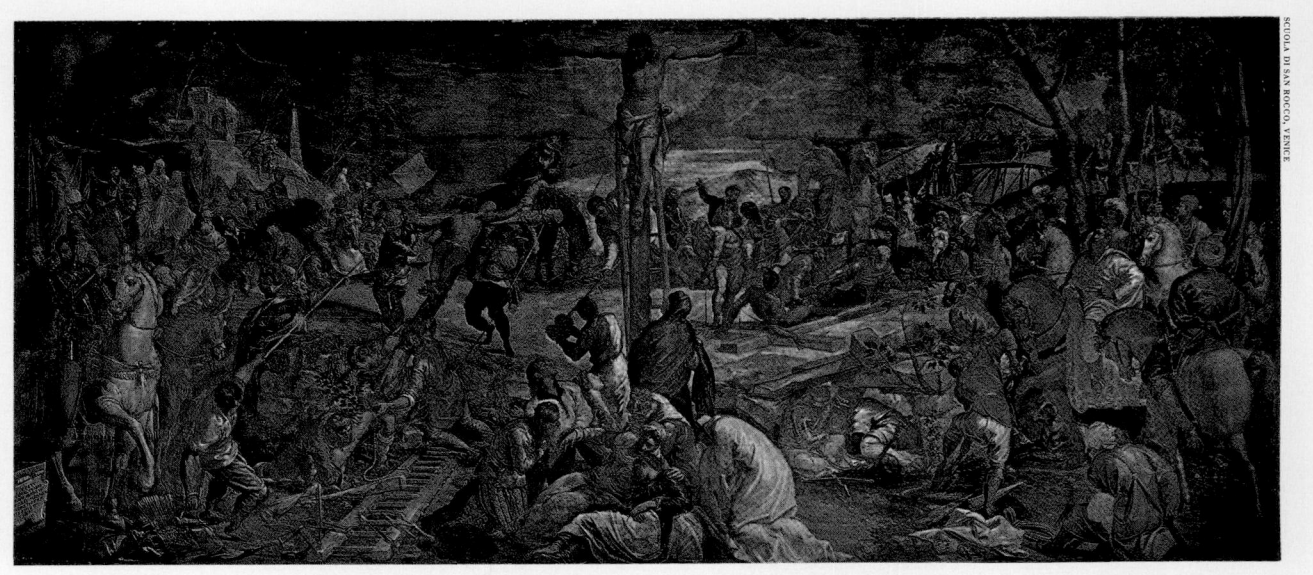

Tintoretto: *The Crucifixion*, 1565

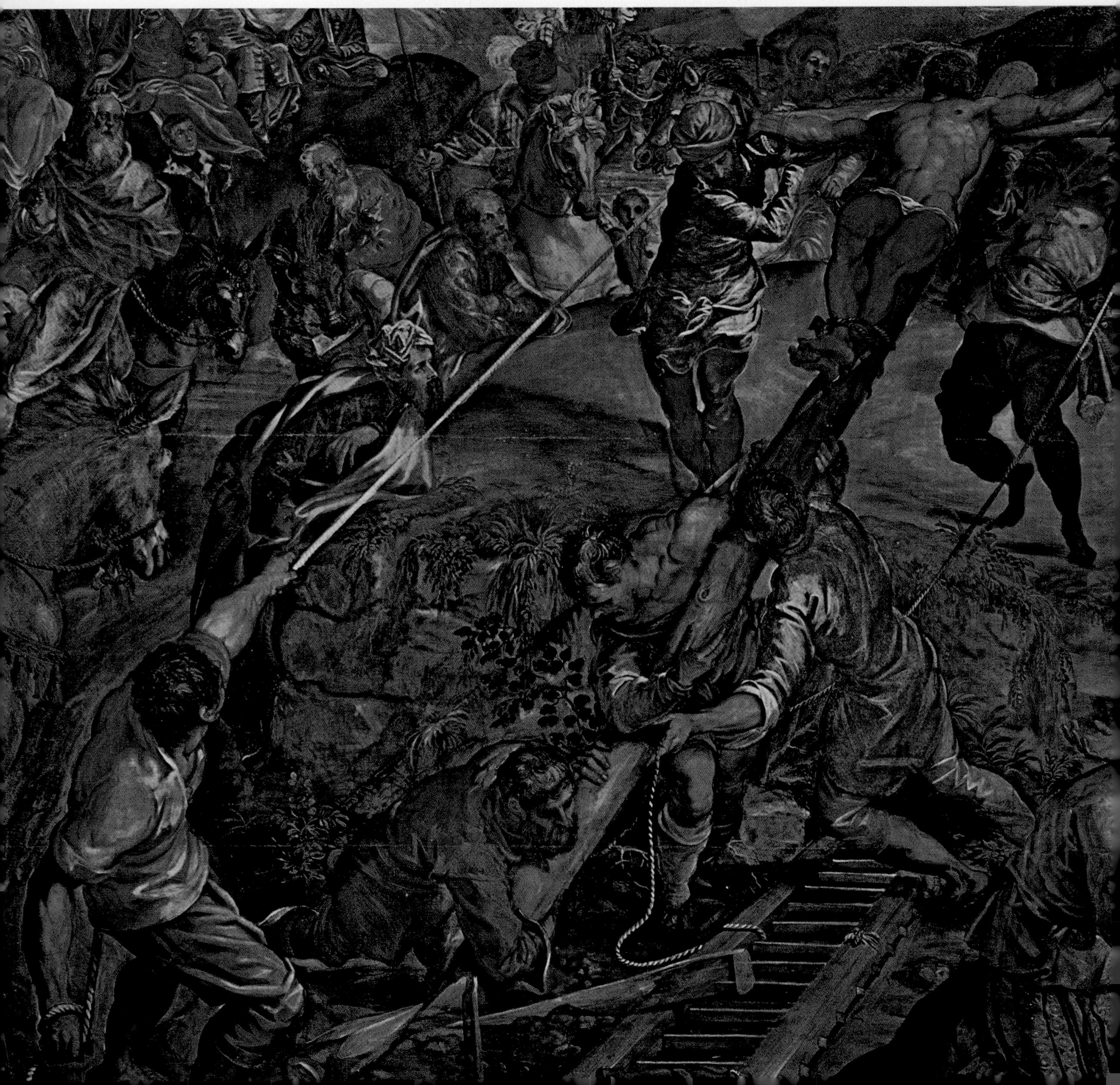

A bold and ardent man, Tintoretto was called *Il Furioso* for the violent way in which he spread paint on a surface and impatiently pushed pigment into place. Tintoretto was intensely attracted to the spiritual world and painted scores of Biblical scenes that were filled with dramatic tumult, light and color. His greatest masterpieces are the monumental paintings in the Scuola of San Rocco, a building owned by a fraternal and religious society. *The Crucifixion* at the left stretches some 40 feet, filling an entire wall in the Scuola's dining room. The central figure

of Christ, surrounded by an aura of unearthly light, dominates the picture, but Tintoretto has filled the scene with a multitude of other activities. In the detail below, the workmen busily lifting the thief's cross at the left, the two soldiers shooting dice for Christ's robe in the hollow at the lower right, seem uncaring of His fate. Only Mary, huddled desolately at the foot of the Cross, her arms outstretched, has begun to grieve. By including these subordinate scenes, Tintoretto has given the agonizing event the sense of a vivid human drama.

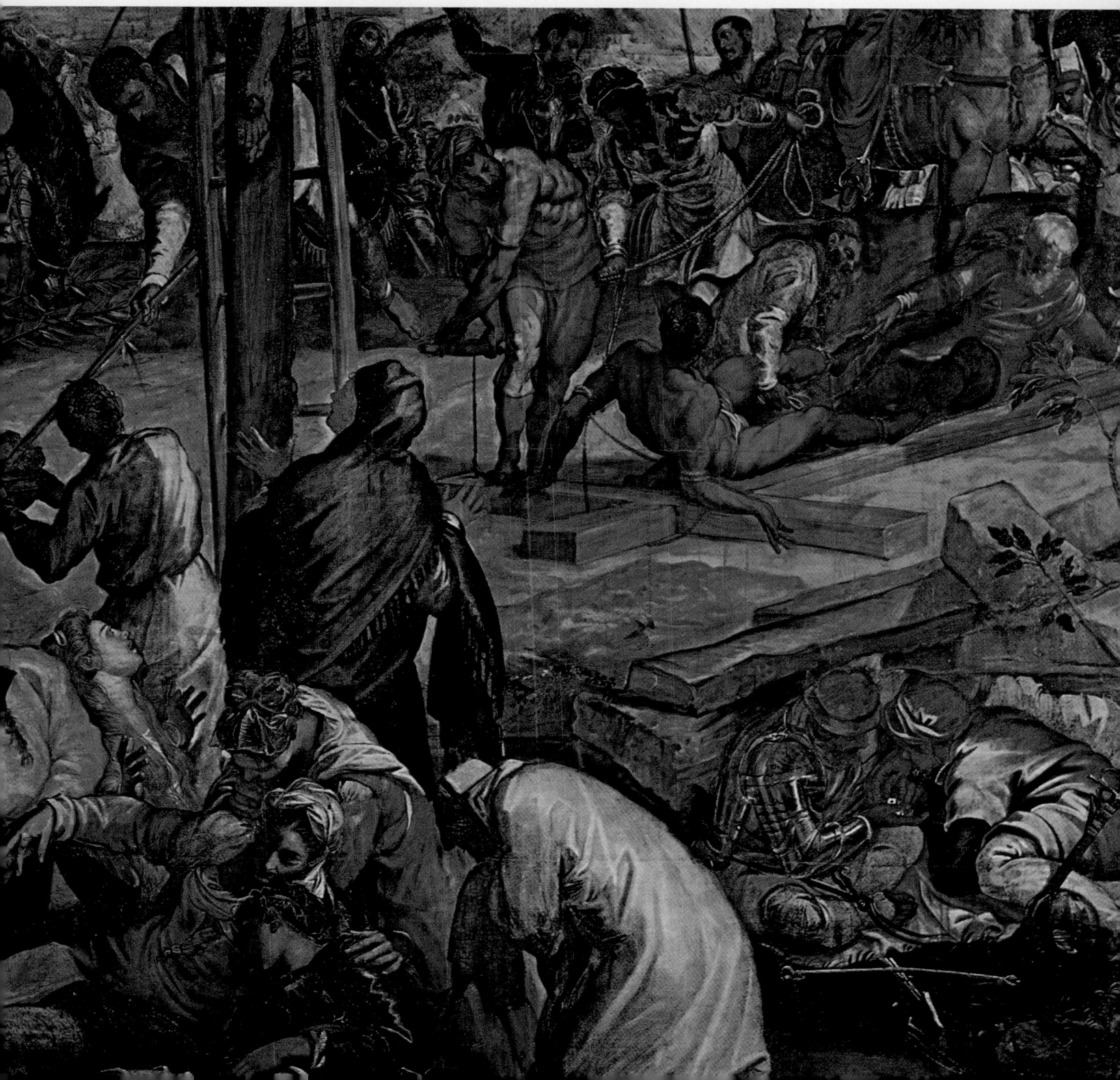

This painting, one of Tintoretto's last major works, imaginatively depicts a miraculous episode from the Old Testament, the gathering of the manna that rained from heaven. The Bible tells how this blessed food—which speckles the center foreground—fell like the dew to feed the Israelites during their 40-year search for the Promised Land.

Tintoretto here displays a concept of pictorial space that is dramatically different from Renaissance convention. Rather than place the primary action in the foreground of the scene and subordinate everything else to it, Tintoretto spread his narrative across the picture's surface. Moses, leader of the Exodus, is in the lower right-hand corner, his importance heightened by the dazzling aura around his head. As for the event that is commemorated in the painting, only a few of the many Israelites are occupied in gathering up the manna. They attract little more attention than the seamstress in the center, the washerwoman behind her, the cobbler at far left, the spinster in center background or the donkey driver in the right background. By thus diverting the viewer's eye to all parts of the canvas, Tintoretto creates a scene that surges with movement, energy and color.

Tintoretto: *The Gathering of Manna*, 1594

Venice's darling among painters, though not a native, played courtier to her majesty. He was Paolo Cagliari, called Veronese because he came from Verona.

Veronese was an exuberant young artist when he arrived in Venice, and he took enormous delight in the round of stately processions and sumptuous banquets that were such regular features of the city's life. His vivid pictorial imagination must have thrilled to the elegant pageantry, the rich fabrics and the dazzling color of Venetian public ceremonies, and even when he was depicting the most solemn Biblical episodes he robed his figures in 16th Century garb. The theme of Veronese's *Marriage at Cana (right)* is Christ's first miracle, the changing of water into wine, but its treatment is pure Venice. Set in a marble *palazzo,* the scene is filled with the sound of music, the taste of wine and the feel of heavy brocades and plush velvets. The traditional view of the Venetians as an especially pleasure-loving, sensual people is nowhere better revealed than in Veronese's celebration of physical details and sensations in this picture and in others, both religious and mythological, that he painted.

Veronese did not invent the big, pageant-filled, narrative scene—the tradition goes back to Gentile Bellini and Carpaccio *(pages 20-21)*—but he brought to it a remarkable technical skill. If his paintings lack something in the way of emotion, they miss nothing of the artist's hand and eye. Among many later painters who were vastly influenced by his work was the great 19th Century colorist Eugène Delacroix, who wrote about Veronese: "There is one man who paints brightly without violent contrasts, who paints the open air, which we have repeatedly been told is impossible. . . . In my opinion he is probably the only one to have caught the whole secret of nature. Without precisely imitating his manner, one can pass along many paths on which he has placed veritable torches."

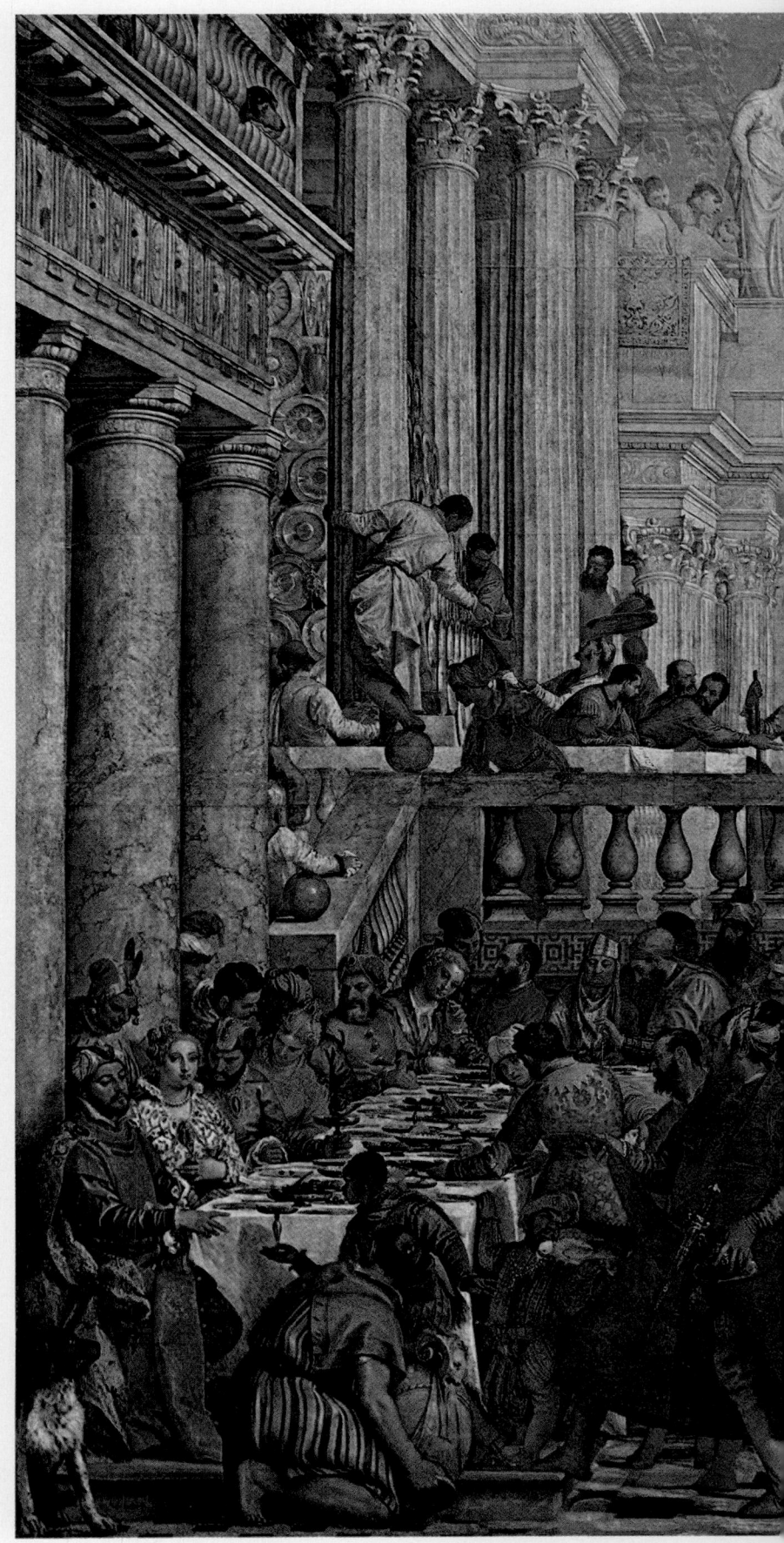

Veronese: *The Marriage at Cana*, 1562-1563

VIII

Genius Undimmed by Age

Veronese's proclivity for dressing
Biblical characters in contemporary
dress is nowhere better seen than in
this picture of the finding of Moses:
Pharaoh's daughter, richly gowned in
the fashion of 16th Century Venice, is
a voluptuous blonde. Even the setting
has been transferred from Egypt to an
Italianate landscape. And the chief
character, Moses, is almost lost in the
crowded scene. It was evidently not
Veronese's aim to re-create history
authentically, but to paint a vivid,
colorful feast for the eye.

Veronese: *Moses Saved from the
Waters of the Nile*, c. 1570

By 1553 Sansovino's library on the Piazzetta San Marco, one of the
glories of Venetian architecture, was just about completed. The vaulted
roof that had collapsed so disastrously eight years before had been restored, and in the library's Great Hall the concave ceiling was ready for
decoration. The Venetian government decided to cover it with a series of
circular allegorical paintings, and assigned seven artists to the project. As
an added inducement, a gold chain was to be awarded to the painter it
deemed to have done the best work. When the ceiling decoration was
finished in 1556, the chain went to a young painter only recently arrived in Venice, Paolo Caliari of Verona, nicknamed Veronese. He
had come to the city to work with the painter Giovanni Battista Ponchino on the ceiling decorations for the meeting hall of the Council of
Ten in the Doge's Palace. Subsequently, at the invitation of the prior
of the Church of St. Sebastian, he had decorated the ceiling of the
sacristy and the nave of the church, and the commission at Sansovino's library had followed.

Veronese's gold chain won him a flood of new commissions and a host
of new friends, among them Titian. Affable, lively, fond of creature
comforts, Veronese seemed to have only one vice—a liking for extravagant clothes. His paintings are reflections of himself. The people in them
are charming, gorgeously dressed, polite and impersonal. Even when the
subject is grim or sad, as in a martyrdom or a Crucifixion, Veronese's
treatment of the event is dispassionate. The sense of life and nature and
light that permeates so much Venetian painting is especially strong in
Veronese's work, but in him these qualities became theatrical, curiously
artificial. Like Venice itself at this point in its history, Veronese's world is
not quite real.

The Venetian republic, once so powerful, was now in decline. Its
mainland possessions were steadily dwindling, and a treaty with the
Turks in 1540, ending a decade of conflict, lost Venice still more of its
possessions in the Levant. The treasury was drained and trade was falling
off, undermined by the galleons of Spain and Portugal, which now
rounded Africa and crossed the Atlantic to bring home Eastern goods and

the wealth of the New World. The Venetian government, never noted for its democracy, was growing more and more autocratic. The Council of Ten, formed in the 14th Century to deal swiftly with anything that threatened the safety of the state—treachery, corruption, subversion—was itself now considered untrustworthy. Many of its powers had been delegated to a Council of Three Inquisitors who were, if possible, even more feared and hated than the Ten.

Yet on the surface Venice was gay and glamorous, never more opulent nor more ostentatious. The city was like a magnificent piece of make-believe. Something of the same quality marked the works of Veronese. His gigantic paintings of feasts, produced for the dining halls of convents and monasteries—*The Feast in the House of Simon the Pharisee, The Feast in the House of Levi, The Marriage at Cana*—hardly seem like religious paintings at all. It is faintly comical to think of a sober row of monks sitting beneath one of these splendid scenes, eating in silent meditation and looking up at these vistas of marble columns and festive banqueters, with their golden vessels, musicians, dwarfs and dogs. Lodovico Dolce once took Albrecht Dürer to task for painting Christ and the Virgin Mary in German clothing instead of Biblical robes. Apparently the question of historical accuracy never entered Veronese's head. His Biblical women blaze with jewels, and their blonde hair is done in the latest mode. In *The Marriage at Cana,* at which Christ turned water into wine, the wedding party includes Queen Eleanor of France, Sultan Suleiman and that late lover of feasting, Pietro Aretino *(pages 166-167).*

On one occasion one of Veronese's banquet scenes got him into trouble. The Roman Catholic Church objected to certain details in *The Feast in the House of Simon,* painted for the Dominican convent of St. John and St. Paul. The painting showed a man with a nosebleed, two German soldiers with halberds, a jester with a parrot, a man picking his teeth and "similar vulgarities" offensive to Rome. Veronese was haled before a tribunal of the Inquisition. "Do you not know," his interrogator cried "that in Germany and in other places infected with heresy it is customary with various pictures full of scurrilousness and similar inventions to mock, vituperate, and scorn the things of the Holy Catholic Church in order to teach bad doctrines to foolish and ignorant people?" Veronese argued that he had lots of space to fill, and had filled it: "I paint pictures as I see fit, and as well as my talent permits." He added that in any case the objectionable figures, the soldiers and buffoons, were outside the banqueting room. Church officials were adamant and ordered him to change the painting within three months. Instead, Veronese simply changed its title, satisfying everyone by calling it *The Feast in the House of Levi,* a banquet attended, according to the Gospel of St. Mark, by "tax collectors and other sinners."

From the nature of Veronese's paintings and the manner in which he worked one gets the impression that his work gave him great joy. The abundance of detail, the superb fabrics, the happy faces of his banqueters, reflect a boundless delight in the painting process and in the Venice and Venetians that he painted. Moreover, he worked with exceptional speed. The huge *Marriage at Cana,* some 20 feet high and 30 feet long, was completed in less than 15 months—a splendid achievement even allowing for the fact that he had several excellent assistants, chief among them his brother Benedetto. Nor does Veronese seem to have cared much about money. Although he was fairly well off when he died, *The Marriage at Cana* was done for his keep, a cask of wine and a modest 324 ducats.

Late in his career, when he was quite famous, Veronese was asked to replace several allegorical paintings on the ceilings of the Doge's Palace, the previous ones having been destroyed by fire. One of these allegories portrays Venice as the Queen of the World, a buxom blonde, handsomely dressed and surrounded by admirers, but somewhat blowsy—rather like a cross between a monarch and a prosperous bordello-keeper. It was a combination that seemed curiously appropriate. The golden city that Carpaccio had painted as a lively, impromptu pageant had now faded to tones of silver, and its pageantry had lost its spontaneity.

Veronese's keenest competitor for Venetian flavor was his close friend, Jacopo Robusti, the son of a well-to-do dyer of woolen cloth—hence his nickname, *Il Tintoretto,* "the little dyer." Robusti also had a second nickname, *Il Furioso,* descriptive of the furious force with which he worked. Tintoretto threw himself into art as fiercely and joyously as a Viking warrior hurled himself into battle; it was a kind of frenzy with him. He made himself a superb draftsman, for instance, by concentrating on drawing, his weakest point. His workshop was hung with jointed manikins, which he sketched over and over from every conceivable angle until he was able to draw any pose accurately and automatically. He did studies of Michelangelo's statues, using casts that he imported from Rome. He constructed miniature scenes, like model stage sets, which he lit from various angles and in various degrees of intensity, so that he could investigate the effects of light and shade. His researches were even said to have led him to watch anatomical dissections and to observe artists of every kind at their work—carpenters, cabinetmakers, furniture painters, decorative plasterers, even fourth-rate daubers who set up shop in the Piazza San Marco and sold their work for pennies.

Tintoretto snapped up every chance of displaying his work and cared so little for money that he did frescoes and canvases for the cost of materials—or for nothing. By the time he was 27, in 1545, Pietro Aretino was writing to thank him for two pictures "with which you, so young in years, have decorated the walls of my salon in less time than it would have taken most people to think about them." For Venice, such praise was sufficient recommendation. Three years later Tintoretto was commissioned to paint a picture of an episode in the life of St. Mark for the prestigious Brotherhood of St. Mark; and with this painting, *Miracle of the Slave,* his reputation was firmly established.

Tintoretto's speed was a matter of wonder. When in 1564 the Scuola of San Rocco asked four artists to submit sketches for a painting to decorate the ceiling of their new building, Tintoretto unveiled a finished picture. He had taken the measurements of the space, completed the painting and had it secretly installed. When the brotherhood and his rivals objected to such tactics, Tintoretto replied that it was the only way he knew how to work; besides, it gave the purchaser a very precise idea of what he was getting. He added that if the brotherhood did not approve of his approach, it could have the painting as a gift. They not only accepted his offer, but for the next 23 years kept him busy decorating the Scuola's ceilings and walls with a series of paintings of episodes from the life of Christ that are still San Rocco's pride and joy.

Looking at a Tintoretto canvas is a bit like standing in the middle of a hurricane. Many of his creations are enormous, as much as 20 feet high and 40 feet long. Hundreds of figures are swept about by the passionate energy of the artist. The themes of his work were often the same as those of other Venetian painters, but the treatment of the themes was vastly different. Where Titian chose to compose a relatively static *Presentation of the Virgin,* seen in profile, Tintoretto viewed the same incident head on and in daring perspective, lit it dramatically and imbued it with a sense of violent movement. "Tintoretto," said Carlo Ridolfi, "never did anything as other people did." Vasari noted: "He may be said to possess the most singular, capricious and determined hand, with the boldest, most extravagant and obstinate brain that has ever yet belonged to the domain of art." Where other artists were content to illustrate events, Tintoretto participated in the events he painted—and drew his audience in with him.

For all his tempestuous paintings, Tintoretto's private life was cheerful and serene. Although he treated his fellow painters somewhat shabbily by undercutting their prices, he does not seem to have made himself any bitter enemies. Veronese, who happened to be one of the competing artists in the San Rocco competition, continued to be a frequent visitor at Tintoretto's home even after that episode. Similarly, Tintoretto's home life was a conventionally happy one: he was married to a noblewoman, lived in a pleasant house bought with the help of his father-in-law and had eight children. In common with almost every other Venetian artist, he loved music and his house was full of it: he played the lute himself and encouraged his children to play various other instruments. A burly, bearded man with a dry sense of humor, he once poked fun at his own preference for large canvases by advising a prospective customer who wanted a Tintoretto fresco for his garden wall to build the wall to "three Tintorettos" in size. On another occasion, when a group of Venetian dignitaries, watching him at work on some decorations in the Doge's Palace, asked him why he painted so fast when other artists were slow and methodical, Tintoretto growled, "Because they haven't got so many pests around to drive them crazy."

Much of the power of Tintoretto's canvases comes from the bold way in which he used his brush. As Venetian painters became more familiar with oil paint, they discovered that the manner in which the paint was applied gave them an added means of expression: the record of a painter's brushstrokes upon the canvas underscored his intent. The Venetians were the first painters to let their brushwork show, to exploit the manner in which they put paint to canvas. Titian built up layer upon layer of transparent oil colors to create the illusion of real flesh; Veronese used the texture of oil colors to simulate damask and satin and fur; Tintoretto diluted his oil colors so that they flowed easily across large areas of canvas, allowing him to create forms almost as fast as he thought of them.

But if the paintings of Tintoretto and Veronese represented Venice's growing skill in the use of oil, they were also representative of changing tastes in art—and indeed contributed to that change. Patrons were demanding a more expressive kind of art, less ideal, more emotional; Italian art was moving away from the ordered, classical style toward a more emotional, expressive style, toward Mannerism and the Baroque.

Inevitably, Titian's work changed too. But instead of following fashion the aged painter traveled a fresh path, one dictated by his own genius. In 1566 Vasari visited Venice and went to call on Titian "as one who was his friend." He found him "still with brushes in his hand and painting busily." Titian was then 78 years old. Vasari, however, cast some doubt on the quality of what he saw. "It would have been well," he wrote, "if he had worked for his amusement alone during these latter years, that he might not diminish the reputation gained in his best days by works of inferior merit, done at a period of life when nature tends inevitably toward decline, and consequent imperfection." Others were less gentle in their remarks. "All say he can no longer see what he is doing," wrote the art dealer Niccolò Stoppio, "and that his hand trembles so much that he cannot finish anything, but leaves it to his assistants." A bit of rather untrustworthy gossip, passed on in the following century by the German painter, Joachim von Sandrart, had it that Titian in his old age liked to alter his earlier paintings, and that his assistants, to prevent what they considered damaging changes, took to mixing his colors with olive oil so that the alterations could be washed off in his absence.

Certainly, Titian's work during the last decade or so of his life must have looked slapdash and unfinished to many of his contemporaries, especially to those who admired minute and careful detail. The colors are softer, the outlines blurred, the earlier precision has been replaced by brushstrokes that only resolve themselves into definite forms when one steps back from the canvas and views it at a distance. The *Martyrdom of St. Lawrence* that Titian sent to Philip II in 1567, for instance, is a far cry from an earlier painting of the same subject, almost identical in composition. The earlier version is like some stately ritual, while the later one is full of agony and motion. There are similar differences between the smoothly finished *Christ Crowned with Thorns,* painted in the 1540s for the Church of Santa Maria delle Grazie in Milan, and a violent and passionate *Christ Crowned with Thorns* painted some 20 years later *(page 161)*.

Tintoretto, when he saw this latter painting, understood exactly what the old man was trying to do and was so struck by it that he begged Titian to give it to him and kept it on display in his own studio. More and more, Titian was being possessed by the medium in which he worked. Instead of pursuing form, he now surrendered himself almost wholly to the color and texture of the paint. Light and shadow, once used to heighten the illusion of reality, now became valuable for themselves alone, as interplays of color. Brushstrokes that had formerly been smoothed away were now left to ripple the surface of the canvas, lending tension and excitement to the painting. For at least three centuries painters were to embrace these innovations of Titian's last years. Rembrandt and Rubens, plying their brushes through thick impasto; Renoir and Van Gogh, laying on patches of light and shadow to create the impression of forms, were manipulating paint in ways first investigated by Titian.

Perhaps Titian's explorations were indeed the by-product of old age; perhaps the change in his art was born of necessity, forced upon him by wavering hands and dimming eyes. Nevertheless, the accomplishment remains. Titian's late paintings surmount his physical weaknesses and even make use of them. As a colorist, he turned naturally to color to create form when he could no longer depend upon the steadiness of his line; as an oil painter, he found it equally natural to exploit the uneven texture of oil in pursuit of an artistic end. One young painter, Palma Giovane, saw Titian at work during this late period and set down an account of his method. From Palma's description it is clear that while the infirmities of age may have helped to drive Titian in a new direction, he was also moving that way by conscious intent:

> Titian began his pictures with a mass of color which served as a bed or foundation for what he wished to express. . . . With four strokes he was capable of indicating a magnificent figure. . . . After he had thus applied this important foundation, he turned the pictures to the wall and left them, without looking at them, sometimes for months. When he afterwards returned to them . . . he brought the skeleton of his figures to the highest degree of perfection, and while one picture was drying, he turned to another. . . . He gave the last touch to his pictures by adjusting with his fingers the transitions from the highest lights to the halftones, or he would apply with his fingers a spot of black in one corner or heighten with a dab of red, like a drop of blood, the liveliness of the surface; and thus he gradually brought his figures to completion. In the last stages of the work, he painted more with his fingers than with the brush.

During the last decade of his life, Titian withdrew from public affairs and prepared for the end. In 1567 he asked the Venetian government to transfer his broker's license in the Fondaco to his son Orazio, and four years later he also turned over to Orazio the income from his Milanese pension. In November 1571, when the government asked him for a commemorative painting, celebrating the month-old victory of the Spanish and Venetian fleets over the Turks at the battle of Lepanto, Titian turned the assignment over to an assistant. To add to his sadness during this decade, the second member of the old Triumvirate died. Jacopo Sansovino, at 93, lay down to rest one day in 1570 and never again arose. "He felt no kind of illness," says Vasari, "but. . .felt himself becoming weaker, and requested to have the Sacraments of the Church administered to him; this having been done, . . . Sansovino departed. . . ." He had been, says Vasari, an indomitable old man, "strong and healthy even in his 93rd year, and able to see the smallest object at whatever distance without glasses, even then. . . . His digestion was so good that he could eat all things; during the summer he lived almost entirely on fruits, and in the extremity of his age he would frequently eat three cucumbers and half a lemon at one time."

Titian had now outlived all his dearest friends, but he was still not too old to strike one last bargain. As in the past he had offered to exchange paintings for certain material favors, such as the loan of a suit of armor, so he now offered to paint a *Pietà* for the Franciscan brothers of the Frari church in return for a burial place in their chapel. The brothers agreed, but Titian, although he began the *Pietà,* did not live to complete it. In the summer of 1575 plague broke out in Venice and raged through the city for over a year, killing more than one quarter of the population. Every day the black gondolas carried their grim cargoes from the city under skies clouded by the smoke from pyres of infected clothing and furniture. Toward the end of the epidemic, on August 27, 1576, Titian himself died, probably of the plague, although curiously no cause of death was recorded. Perhaps in the terror and stress of the times, municipal documents were ill-kept or mislaid. All that is known for certain is that, despite the unfinished *Pietà,* he was buried in the chapel of the Frari church. The guild of Venetian painters wanted to give him a splendid funeral, but the Venetian authorities, terrified of the dangers of public gatherings, forbade it. Moreover, before anyone could lift a hand to prevent it, Titian's pleasant house in the Biri Grande was broken into and looted.

Fate dealt unkindly with Titian in the years immediately after his death. His good son Orazio followed him into the grave within a few days, while the bad son, Pomponio, lived on for at least 18 years, selling off all his father's pictures and possessions to support himself and ending his life as an obscure pauper. Within a year of Titian's passing, all his paintings in the Doge's Palace were destroyed by fire, and other fires, plus wars, time and clumsy restorations played havoc with many of the rest. But there is ample proof of his genius in the work that remains as well as his profound influence on other painters. Generations of artists, copying the intoxicating colors of a *Bacchanal* or responding to the emotional power of a *Christ Crowned with Thorns,* have recognized and followed his authority. It would be hard to imagine what the next 300 years of painting after Titian would have been like without him. The depth of his perceptions and naturalness of his vision helped to create an art that was independent of, and superior to, an art concerned solely with reality. He gave a new meaning and a new dimension to the concept of beauty in art. Titian was one of those rare seminal spirits after whose passing the world is never again quite the same.

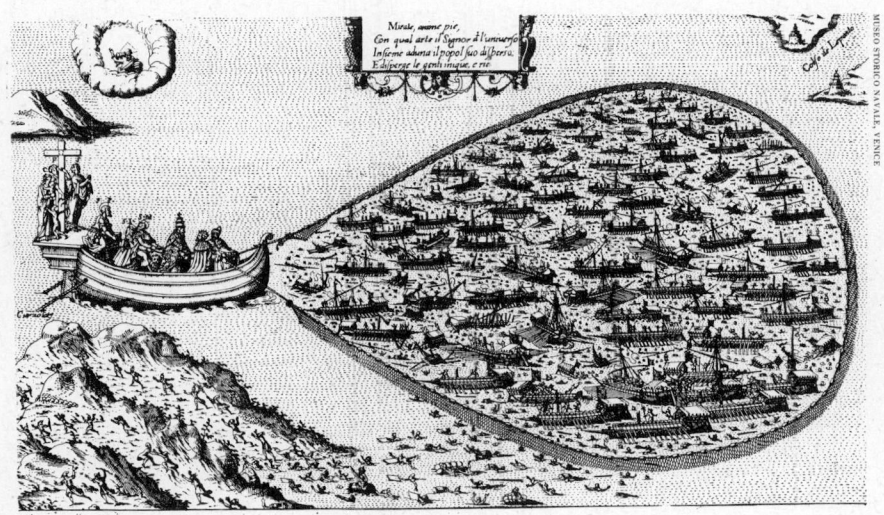

The Turkish fleet is trapped like fish in a net in this contemporary allegorical engraving commemorating the battle of Lepanto in 1571. It was there that allied Christian forces ended forever the Sultan's control of the Mediterranean. The three architects of the victory appear in the bow of the boat: Pope Pius V, Philip II and the Venetian Doge. Behind them stand three saints and, under the cross, three angels. God Himself is shown overseeing the round-up from the cloud at upper left; the caption at the top reads, in translation, "Pious souls admire how artfully the master of the Universe unites his peoples and confounds the wicked."

Venice Revisited

The heartbeat of Venetian art all but stopped with the deaths of Titian, Tintoretto and Veronese. No masters of their stature emerged to replace them and Venice went into decline, unable to dominate trade as she once had. Furthermore, an art-minded papacy attracted the painters with new ideas, and early in the 17th Century, Rome became the center of art in Italy.

Gradually, however, the balance shifted back again. Venice—and her great tradition in art—revived with the help of foreign pleasure-seekers, who lavished money and affection on their luxurious playground. So stunning was the comeback that, early in the 18th Century, Venice led all Italy in one final display of creativity before Paris assumed leadership of the art world.

The genius of this late but vibrant period was Giovanni Battista Tiepolo, the most sought-after decorative artist in all Europe. Three other outstanding men who contributed to Venice's renewed artistic vigor were the painters of city views, Canaletto and Guardi, and a chronicler of daily life, Pietro Longhi.

It is perhaps a kind of poetic justice that these last three, all native-born Venetians, should have been the ones to memorialize their city and its inhabitants in the last decades of her independence. For in 1797, Napoleon Bonaparte, sweeping triumphantly through Europe, put an end to the 1,300-year-old republic and, with it, ended Venetian art as a unified tradition.

Tiepolo won fame and fortune throughout Europe for his highly theatrical fresco decorations for churches, palaces and private villas. The soaring, light-filled scene shown here portrays the legendary Aeneas being escorted heavenward by winged Victory and lion-led Merit. There Aeneas is to meet his mother Venus and receive the helmet that will admit him into the Temple of Immortality. The fresco was one of three that Spain's Charles III commissioned for ceilings in the new Royal Palace at Madrid.

Giovanni Battista Tiepolo:
Apotheosis of Aeneas, 1764-1766

Pietro Longhi was 18th Century Venice's most popular painter of genre pictures, many of which chronicle the aristocracy in idle search of amusement. The painting at left below depicts a typical gathering in a notorious gambling hall, the Ridotto, where the rich and the riffraff mingled. At right is a record of a fashionable happening—the first exhibit in Venice of the exotic rhinoceros, at a carnival in 1751. Although he was never considered a great

Pietro Longhi: *The Ridotto,* c. 1757

painter, Longhi charmed his contemporaries with amiable, pictorial small talk. Having found the formula for success, he changed his style very little, even repeating the same figure in different paintings, like the bare-shouldered woman in the background of both of these pictures. To his credit, Longhi was an unflagging observer of the most casual action and detail, and as a result, his work presents a vivid account of the everyday life of his times.

Pietro Longhi: *The Rhinoceros,* 1751

Francesco Guardi: *View of Santa Maria della Salute,* c. 1775-1780

A city of visual delights, Venice became an irresistible subject for artists who arrived from northern Europe in the 17th Century. Spurred by their enthusiasm and precise renderings, Italian artists went on to evolve a new type of scenic painting devoted to city views, called *vedute.* Their beguiling pictures enjoyed a special vogue as travel souvenirs well into the following century when wealthy tourists, especially the British, flocked to Venice.

View painting reached its peak in the work of Canaletto and Guardi. Guardi, less interested in exact detail than Canaletto, created lively interpretations of the world around him. Imaginative and capricious, he bathed freely drawn boats and buildings in a soft, liquid atmosphere that lent an element of fantasy to familiar sights *(left).* Canaletto, in contrast, flooded his scenes of Venice with the brilliant light of high noon. Trained in youth as a stage designer, he was at ease with the detail and perspective needed to record landmarks like the Doge's Palace with the crisp focus of a camera lens *(following pages).*

The homage both painters paid to Venice in her final years as a city-state brightened her sunset glow and set a felicitous pattern. For in years to come, foreign artists such as Auguste Renoir, J.M.W. Turner, James Whistler, Oskar Kokoschka and Raoul Dufy—to name just a few— joined Guardi and Canaletto in capturing the sparkle and splendor of a unique subject.

Canaletto: *View of the Ducal Palace and the Piazzetta San Marco*, date unknown

How the Venetians Painted

Venetian painters at the time of Titian used a different technique from that of later painters. The modern artist usually paints the forms of his composition directly on the canvas with color mixed on his palette. This is essentially a one-step method, although the artist may paint over a preliminary sketch. By contrast, the Venetian painter used a two-step method. First, he defined the forms of his composition in monochrome, and only after that was completed did he apply color. When he applied color, he did so in translucent layers called glazes.

The Venetian artist knew the use of various kinds of paint, which consisted, then as now, of powdered color (pigment) mixed into a liquid (vehicle) capable of drying hard and thus holding the pigment in place. The craft of painting was well advanced by Titian's time. For centuries before him artists had been trying various drying oils and temperas, sometimes in combination and with varying success; some oils darkened or turned yellow; others cracked or took too long to dry. Tempera, usually made with egg yolk and water, was favored for its permanence. But because it dried rapidly it proved impossible to blend on the painting surface. Forms could be modeled and given dimension only with a laborious, hatching brushstroke. Tempera paintings also tended to have a flat, opaque appearance.

In about 1400 Flemish artists, notably the van Eyck brothers, developed a method that used both tempera and oil vehicles. The work was divided into two basic stages: an opaque underpainting and overlayers of translucent color glazes. In the first stage all forms were carefully modeled in pale tones of gray. In the glazing stage pure colors—pigments not mixed with white—were combined with a clear, varnishlike oil medium of great gloss and applied over the underpainting, allowing the forms to show through.

A picture painted in this fashion has extraordinary luminosity and depth. Light falling on it passes through the colored glazes, reflects from the opaque underpaint and returns to the eye of the viewer. Thus, colors are seen by transmitted light, as in a stained-glass window.

The northern European art patrons liked small, delicate pictures, but the wealth and power of 16th Century Venice called for grandiose paintings. In meeting these demands Venetian artists, beginning with Titian, recognized the advantages of the Flemish technique of applying transparent glazes of color on top of forms modeled in monochrome. The same method that had produced small wooden panel paintings of jewel-like quality now was modified to achieve dramatic, harmoniously colored paintings of monumental proportion.

Wood panels proved too confining so the Venetians turned to canvas, which could be woven to large sizes and joined to provide the even larger surfaces that their grand vision sometimes required. (The rougher texture of the canvas also provided grip for their freely applied paint.) They first prepared the canvas with a white ground. The ground used by the Venetians was essentially composed of gypsum mixed with glue. But instead of working directly on this glaring surface they started by brushing on a thin film of brown or reddish-black paint. On this dark ground they began work with white, loosely at first, then more fully developing the various forms of the composition. For this underpainting they favored a mixed vehicle made of a tempera and oil emulsion, which set quickly and offered great flexibility, from thickly painted light areas to veil-like washes for the shadows. Color glazing and detail work came at a later stage.

To demonstrate the Venetian method of painting for this book, artist Jane Flora was commissioned to copy a portion of Titian's *Madonna with the Rabbit* shown on page 95. The intention is not to demonstrate exactly how Titian painted this picture, but to illustrate the main steps of work in a highly simplified demonstration of fundamentals. The four steps illustrated in the panels at the left are:

1. *The toned ground.* The artist first stains the white of the canvas brownish red and then sketches in the main outlines with a darker, thinned-out paint.

2. *Underpainting.* Dimension is achieved by modeling the forms in monochrome. In some dark areas the background is allowed to show through; in others, black is mixed with white to deepen the gray.

3. *Glazing.* After the underpainting has thoroughly dried, the basic colors of the composition are applied in transparent glazes of color mixed with oil. The color is boldly stated at this stage, without great concern for the overall harmonies.

4. *Finishing.* The forms and details are now fully developed by delicate heightening with white and with additional color glazes brushed on, or perhaps rubbed in with the finger. Tradition has it that Titian once exclaimed, "Glazes, thirty or forty!" in explanation of how he achieved his color effects. True or not, the implication of the story is correct. He certainly returned again and again to his paintings, modifying and adjusting the tonal balances and color relationships.

Artists of Titian's Era

VENICE
DOMENICO VENEZIANO 1400-1461
JACOPO BELLINI c.1400-c.1470
ANTONELLO DA MESSINA 1414/30-1479/93
ANTONIO VIVARINI c.1415-1476/84
GENTILE BELLINI 1429-1507
GIOVANNI BELLINI c.1430-1516
BARTOLOMMEO VIVARINI c.1432-c.1499
CARLO CRIVELLI 1435/40-c.1493
ALVISE VIVARINI 1445/46-1503/05
GIOVANNI D'ALEMAGNA fl.1450-60
VITTORE CARPACCIO c.1450-c.1525
GIORGIONE DA CASTELFRANCO 1477/78-1510
PALMA VECCHIO c.1480-1528
LORENZO LOTTO 1480-1556
SEBASTIANO DEL PIOMBO c.1485-1547
JACOPO SANSOVINO 1486-1570
TITIAN (TIZIANO VECELLIO) c.1488-1576
SEBASTIANO ZUCCATO ?-1527
GIOVANNI-BATTISTA PONCHINO c.1500-1570
JACOPO BASSANO 1515/16-1592
TINTORETTO (JACOPO ROBUSTI) 1518-1594
ORAZIO VECELLIO c.1525-1576
GIAN MARIA VERDEZZOTTI 1525-1600
GIOVANE (JACOPO PALMA) 1544-1628

FLORENCE
MASACCIO (TOMMASO GUIDI) 1401-1428
FRA FILIPPO LIPPI c.1406-1469
ANDREA DEL CASTAGNO c.1410-1457
BENOZZO GOZZOLI 1420-1497
ALESSO BALDOVINETTI 1425-1499
ANTONIO POLLAIUOLO 1433-1498
VERROCCHIO (ANDREA CIONI) 1436-1488
SANDRO BOTTICELLI 1444-1510
DOMENICO GHIRLANDAIO 1449-1494
LEONARDO DA VINCI 1452-1519
PIERO DI COSIMO 1462-1521
FRA BARTOLOMMEO 1472-1517
MICHELANGELO BUONARROTI 1475-1564
ANDREA DEL SARTO 1486/88-1530
BARTOLOMMEO BANDINELLI 1493-1560
JACOPO DA PONTORMO 1493/94-1557/58
ROSSO FIORENTINO 1494-1541
NICCOLO TRIBOLO 1500-1550
BENVENUTO CELLINI 1500-1571
ANGELO BRONZINO 1503-1572

SIENA
BALDASSARE OF SIENA fl.15th C.
MATTEO DI GIOVANNI 1403-1433
GIOVANNI DI PAOLO c.1403-c.1482
VECCHIETTA (LORENZO DI PIETRO) c.1412-1480
FRANCESCO DI GIORGIO 1439-1502
NEROCCIO DI LANDI 1447-1500
SODOMA (GIOVANNI ANTONIO BAZZI SODOMA) 1477-1549
BECCAFUMI (DOMENICO DI PACE) c.1486-1551

CENTRAL ITALY
PIERO DELLA FRANCESCA 1410/20-1492
MELOZZO DA FORLI 1438-1494
LUCA SIGNORELLI c.1441-1523
PERUGINO (PIETRO VANNUCCI) 1446/47-1523
BERNARDINO BETTI PINTURICCHIO 1454-1513
MARCANTONIO RAIMONDI c.1480-1527/34
RAPHAËL URBINAS 1483-1520
GIULIO ROMANO 1499-1546
GIORGIO VASARI 1511-1574

NORTHERN ITALY
VINCENZO DE FOPPA 1425/30-1515/16
COSIMO TURA c.1430-1495
ANDREA MANTEGNA 1431-1506
FRANCESCO COSSA c.1435-1477
PELLEGRINO DA SAN DANIELE c.1467-1547
BERNARDINO LUINI c.1475-1532
DOSSO DOSSI c.1479-1542
PORDENONE (JEAN-ANTONIO LICINIO) 1483-1576
DOMENICO CAMPAGNOLA 1484-1550
ALFONSO LOMBARDI 1487/97-1537
LAZZARO DA CORREGGIO fl.1494-1534
ALESSANDRO BONVICINO MORETTO c.1498-1554
PARMIGIANINO (GIROLAMO MAZZOLA) 1503-1540
GUGLIELMO DELLA PORTA ?-1577
GIOVANNI-BATTISTA MORONI c.1525-1578
VERONESE (PAOLO CALIARI) 1528-1588

FRANCE
JEAN FOUQUET 1420-1477/81
NICOLAS FROMENT fl. 1450-1490
MAÎTRE DE MOULINS fl.1480-1500
JEAN CLOUET c.1486-1541

GERMANY
HANS MULTSCHER c.1400-1467
KONRAD WITZ (SWISS) 1400/10-1444/46
STEPHAN LOCHNER fl.1410-1451
MEISTER FRANCKE fl.1417-1435
LUCAS MOSER fl.1431-1440
FRIEDRICH PACHER 1435/40-c.1508
LUDWIG SCHONGAUER, THE ELDER c.1440-c.1492
HANS HOLBEIN, THE ELDER 1460/65-1524
MATTHIAS GRÜNEWALD 1470/80-1528
ALBRECHT DÜRER 1471-1528
LUCAS CRANACH 1472-1553
HANS BURGKMAIR c.1473-1553/59
ALBRECHT ALTDORFER c.1480-1538
BALDUNG GRIEN 1484/85-1545
HANS HOLBEIN, THE YOUNGER c.1497-1543

FLANDERS
ROGER VAN DER WEYDEN 1399/1400-1464
HUGO VAN DER GOES c.1420-1482
HANS MEMLING 1425/40-1494
PETRUS CHRISTUS fl.1440-1445
GERARD DAVID c.1450-1523
QUENTIN MASSYS c.1466-c.1530
JAN VAN SCOREL 1475-1562
MABUSE (JAN GOSSAERT) c.1478-1533
JOACHIM PATINIR c.1485-1524
PIETER BRUEGEL c.1528-1569
JOHANN CALKER fl.1530-1590
EGIDIUS SADELER 1570-1629
PETER PAUL RUBENS 1577-1640

HOLLAND
DIRCK BOUTS c.1400-1475
HIERONYMUS BOSCH c.1450/60-1516
LUCAS VAN LEYDEN 1494-1538
MARTEN-JACOBSZ. VAN VEEN HEEMSKERK 1498-1574
JACOB SEISENEGGER 1505-1567
PIETER AERTSEN 1507/08-1575
ANTONIO MOR (MORO) 1519-1575

SPAIN
BARTOLOMÉ BERMEJO fl.1474-1495
LUIS MORALES c.1509-1586
ALONSO SANCHEZ COELLO 1515-1590
EL GRECO (DOMENIKOS THEOTOCOPOULOS) 1540/50-1614

Titian's predecessors, contemporaries and successors are grouped here in chronological order according to school (Venice, Florence, etc.) or country. The bands correspond to the life-spans of the artists or, where this information is unknown, to the approximate periods when they flourished (indicated by the abbreviation "fl.").

Bibliography   *Paperback

TITIAN—HIS LIFE AND WORKS

Crowe, J. A. and G. B. Cavalcaselle, *Titian: His Life and Times* (2 vols.). John Murray, London, 1877.
Gronau, Georg, *Titian*. Charles Scribner's Sons, 1904.
Morassi, Antonio, *Titian*. New York Graphic Society, 1964.
Northcote, James, *The Life of Titian* (2 vols.). Henry Colburn and Richard Bentley, London, 1830.
Phillips, Claude, *Titian: A Study of His Life and Work*. The Macmillan Company, 1898.
Ricketts, Charles, *Titian*. Methuen & Company, Ltd., London, 1910.
Riggs, Arthur Stanley, *Titian the Magnificent*. The Bobbs-Merrill Company, 1946.
Sutton, Denys, *Titian*. Barnes & Noble, Inc., 1963.
Tietze, Hans, *Titian,* second edition revised. Phaidon Publishers Inc., 1950.
Valcanover, Francesco, *All the Paintings of Titian* (4 vols.). Translated by Sylvia J. Tomalin. Hawthorn Books, Inc., 1960.
Walker, John, *Bellini and Titian at Ferrara: A Study of Styles and Taste*. Phaidon Publishers Inc., 1956.

ON OTHER PAINTERS

Baldass, Ludwig, *Giorgione*. Harry Abrams, Inc., 1965.
Cook, Herbert F., *Giorgione*. George Bell & Sons, London, 1900.
Hendy, Philip and Ludwig Goldscheider, *Giovanni Bellini*. Oxford University Press, 1945.
Lauts, Jan, *Carpaccio, Paintings and Drawings, Complete Edition*. Phaidon Publishers Inc., 1962.
Orliac, Antoine, *Veronese*. Translated by Mary Chamot. Hyperion Press, Paris, 1940.
Osmond, Percy H., *Paolo Veronese, His Career and Work*. Sheldon, London, 1927.
Phillipps, Evelyn March, *Tintoretto*. Methuen & Company, Ltd., London, 1911.
Phillips, Duncan, *The Leadership of Giorgione*. The American Federation of Arts, 1937.
Pignatti, Terisio, *Carpaccio*. Albert Skira, Geneva, 1958.
Richter, George M., *Giorgio da Castelfranco*. The University of Chicago Press, 1937.
Tietze, Hans, *Tintoretto, the Paintings and Drawings*. Phaidon Publishers Inc., 1948.
Vasari, Giorgio, *The Lives of the Painters, Sculptors and Architects* (4 vols.). Translated by A. B. Hinds. E. P. Dutton (Everyman's Library), 1963. (Also paperback, abridged and edited by Betty Burroughs, Simon and Schuster, 1946.)
Wind, Edgar, *Bellini's Feast of the Gods. A Study in Venetian Humanism*. Harvard University Press, 1948.

ON VENICE

Brion, Marcel, *Venice. The Masque of Italy*. Crown Publishers, Inc., 1962.
Brown, Horatio F., *Studies in the History of Venice* (2 vols.). E. P. Dutton and Co., 1907. *Venice, an Historical Sketch of the Republic*. G. P. Putnam's Sons, 1893.
Elston, Roy, *The Traveller's Handbook to Northern & Central Italy Including Rome*. Simpkin, Marshall, Hamilton, Kent & Company, Inc., 1927.
Faure, Gabriel, *Venice*. Translated and adapted by J. H. Denis and J. M. Denis. Essential Books, Inc., 1957.
Fraigneau, André and Michel Déon, *The Venice I Love*. Introduction by Jean Cocteau, translated by Ruth Whipple Fermaud. Tudor Publishing Company, 1957.
Lassaigne, Jacques and others, *Venice*. Albert Skira, Geneva, 1956.
Lucas, E. V., *A Wanderer in Venice*. The Macmillan Company, 1924.
McCarthy, Mary, *Venice Observed*. Reynal & Company, 1956.
Molmenti, Pompeo, *Venice* (6 vols.). Translated by Horatio F. Brown. Istituto Italiano d'Arti Grafiche, Bergamo, 1907.
Muraro, Michelangelo, *Invitation to Venice*. Introduction by Peggy Guggenheim, translated by Isabel Quigly. Trident Press, 1963.
Storti, Amedeo, *You in Venice.* * Ed. A. Storti, Venice, 1965.

CULTURAL AND HISTORICAL BACKGROUND

Armstrong, Edward, *The Emperor Charles V* (2 vols.). The Macmillan Co., 1902.
Brandi, Karl, *The Emperor Charles V*. Translated by C. V. Wedgwood. Alfred A. Knopf, 1939.
Burckhardt, Jacob, *The Civilization of the Renaissance in Italy: An Essay* (2 vols.). Phaidon Publishers Inc., 1952. (Also paperback, 2 vols. Harper Torchbooks, 1958.)
Cartwright, Julia, *Isabella d'Este* (2 vols.). E. P. Dutton and Company, 1932.
Chubb, Thomas Caldecot, *Aretino, Scourge of Princes*. Reynal and Hitchcock, 1940.
Cleugh, James, *The Divine Aretino: A Biography*. Stein and Day, 1966.
Dolce, Lodovico, *Aretin: Or, A Dialogue on Painting*. Glasgow, 1870.
Hutton, Edward, *Pietro Aretino, The Scourge of Princes*. Houghton Mifflin Company, 1923.
Plumb, J. H., *The Italian Renaissance, A Concise History of Its History and Culture.* * Harper & Row, 1965.
Putnam, Samuel (translator), *The Works of Aretino* (2 vols.). Critical and biographical essay by Samuel Putnam. Covici, Friede, Inc., 1933.
Symonds, John Addington, *The Renaissance in Italy* (4 vols.). Modern Library, 1935.
Tyler, Royall, *The Emperor Charles V*. Essential Books, Inc., 1956.

ART—HISTORICAL BACKGROUND

Ackerman, James S., *Palladio.* * Penguin Books, 1966.
Berenson, Bernard, *The Italian Painters of the Renaissance*. Meridian Books, 1964.
Chastel, André, *Italian Art*. Translated by Peter and Linda Murray. Thomas Yoseloff, 1963.
Constable, W. G., *The Painter's Workshop*. Oxford University Press, 1954.
Crowe, J. A. and G. B. Cavalcaselle, *History of Painting in North Italy* (2 vols.). John Murray, London, 1811.
DeWald, Ernest T., *Italian Painting, 1200-1600*. Holt, Rinehart and Winston, 1961.
Ridolfi, Carlo, *Le Maraviglie dell'Arte (della Pittura)*, 2 vols. Grote, Berlin, 1914-1924.
Wittkower, Rudolf, *Art and Architecture in Italy, 1600-1750*. The Pelican History of Art, edited by Nikolaus Pevsner. Pelican Books, 1958.
Wölfflin, Heinrich, *Principles of Art History. The Problem of the Development of Style in Later Art,* seventh edition.* Translated by M. D. Hottinger. Dover Publications, Inc., 1929.

Picture Credits

The sources for the illustrations in this book appear below. Credits for pictures from left to right are separated by semicolons, from top to bottom by dashes.

SLIPCASE: Scala

END PAPERS:
Museo Civico Correr, Venice

CHAPTER 1: 6—Staatliche Museen, Berlin, Gemaeldegalerie. 8—Soprintendenza alle Gallerie, Venice. 11—Museo Civico Correr, Venice. 17—Dmitri Kessel. 18—Eric Schaal. 19—Scala. 20, 21—Scala—Erich Lessing from Magnum. 22, 23—Scala (2).
CHAPTER 2: 24—National Gallery, London. 26, 27—Giuliano Viti (4). 30—Museo Civico Correr, Venice. 33—National Gallery of Art, Washington, D.C. 35—Edwin Smith; Museo Vetrario di Murano, Venice. 37—Dmitri Kessel. 38, 39—Aldo Durazzi. 40, 41—© 1962 by Charles E. Rotkin. 42, 43—Ugo Mulas (2). 44—Aldo Durazzi. 45-49—Museo Civico Correr, Venice (5). 50, 51—Feruzzi.
CHAPTER 3: 52—Giraudon. 54—Museo Civico Correr, Venice (2). 58, 59—The New York Public Library Prints Division. 65—Scala. 66—Scala. 67-69—Erich Lessing from Magnum. 70, 71—Scala. 72, 73—Eric Schaal. 74, 75—Scala.
CHAPTER 4: 76—Erich Lessing from Magnum. 78—Osvaldo Bohm courtesy Museo Civico Correr, Venice. 81—Inge Morath from Magnum. 87—Alinari. 88—Novosti Press Agency courtesy State Hermitage, Leningrad. 89—The Metropolitan Museum of Art. 90-93—Lee Boltin (3). 94—Erich Lessing from Magnum courtesy Kunsthistorisches Museum, Vienna; Lee Boltin. 95—Eric Schaal courtesy Musée du Louvre (2). 96—Aldo Durazzi courtesy Galleria degli Uffizi, Florence

(2). 97-99—Scala (2).
CHAPTER 5: 100—Aldo Durazzi. 102—© The Warburg Institute, London (2). 106—Museo Civico Correr, Venice. 107—The New York Public Library Prints Division from *Omnium Feregentium,* Fernando Bertelli (2). 113—National Gallery, London. 114—Scala. 115—National Gallery, London. 116—Scala. 117—National Gallery of Art, Washington, D.C.—National Gallery of Art, Washington, D.C.; Scala. 118,119—Aldo Durazzi (2).
CHAPTER 6: 120—Lee Boltin. 123—Kunsthistorisches Museum, Vienna. 127—Scala courtesy Gabinetto dei Desegni, Galleria degli Uffizi, Florence. 131-133—© A.C.L. Bruxelles (9). 134, 135—Patrimonio Nacional (5). 136, 137—Andreas Feininger (2); Title page for 1523 Zaragoza Edition of *Cortes's Second Letter to Charles V* courtesy The New York Public Library Prints Division; Erich Lessing from Magnum (2). 138—Kunsthistorisches Museum, Vienna. 139—Reproduced from *The History of the Reign of Charles V,* William Robertson, Volume I, Pub. by W. Strahan; T. Cadell, in the Strand, and J. Balfour at Edinburgh, 1779, page 193 courtesy The New York Public Library Prints Division. 140—Staatliche Museen, Berlin, Gemaeldegalerie; Scala—Erich Lessing from Magnum. 141, 142—Lee Boltin (2). 143-145—Scala (2). 146—Lee Boltin (2). 147—Scala.
CHAPTER 7: 148—Derek Bayes courtesy Fitzwilliam Museum, Cambridge. 153—The Metropolitan Museum of Art. 159—© Bibliothèque Royale, Brussels. 161—Scala. 162, 163—Erich Lessing from Magnum—Scala. 164-167—Scala (2).
CHAPTER 8: 168—Lee Boltin. 175—Museo Storico Navale, Venice. 177—Augusto Meneses. 178, 179—Aldo Durazzi (2). 180, 181—Derek Bayes courtesy The Wallace Collection, London. 182, 183—Scala. 184—Paul Jensen (4).

Acknowledgments

For her help in the production of this book the author and editors particularly wish to thank Olga Raggio, Associate Curator, Renaissance Art, The Metropolitan Museum of Art. They also wish to thank the following people: Director Erwin Maria Auer and Georg J. Kugler, Kunsthistorisches Museum, Vienna; Luisa Becherucci, Director, Galleria degli Uffizi, Florence; Lucia Casanova, Director, Library, Museo Civico Correr, Venice; Gian Battista Rubin De Cervin, Director, Museo Storico Navale, Venice; Gabinetto dei Disegni e Stampe, Galleria degli Uffizi, Florence; Angel Oliveras Guart, Inspector General of Palaces and Museums of the Patrimonio Nacional, Madrid; Ernest Iacono, Italian Cultural Institute, New York; Karl Kup and Elizabeth E. Roth, The New York Public Library; Wilhelm Köhler, Staatliche Museen, Berlin, Gemaeldegalerie; Giovanni Mariacher, Director, Musei Civici, Venice; Amelia Mezzetti, Soprintendenza alle Gallerie, Modena; Musei Civici, Ferrara; Terisio Pignatti, Vice Director, Musei Civici, Venice; Joseph T. Rankin and Naomi Street, Art and Architecture Division, The New York Public Library; Aldo Soppelsa, Museo Vetrario, Murano; Victor B. Sullam, Professor of Economic History, Johns Hopkins University; Piero Torriti, Soprintendenza alle Gallerie delle Marche, Urbino; Germaine Tureau, Chief, Section Commerciale de la Photothèque des Musées, Paris; Don Giuseppe Ungaro, Santa Maria dei Frari, Venice; Francesco Valcanover, Soprintendenza alle Gallerie, Venice; Gastone Ventura, Venice; Fernando Fueytes de Villavicencio, Managing-Counsellor of the Patrimonio Nacional, Madrid.

Index

Index (continued)

Index (continued)

The text for this book was set in photocomposed Bembo. First cut in Europe in 1930, Bembo is named for the Italian Pietro Bembo (1470-1547), an arbiter of literary taste. While it has some resemblance to letters designed in 1470 by Nicholas Jenson, it is largely based on characters cut by Francesco Griffo around 1490.

✠

PRODUCTION STAFF FOR TIME INCORPORATED
John L. Hallenbeck (*Vice President and Director of Production*),
Robert E. Foy and Caroline Ferri
Text photocomposed under the direction of Albert J. Dunn and Arthur J. Dunn